Natural and Man-made Air Pollution

Stephen E. Blewett

with Mary Embree

Ventura, California

Seaview Publishing
P.O. Box 2625
Ventura, CA 93001-2625

What's in the Air:
Natural and Man-made Air Pollution
Is an Environmental Research & Education Project
Seaview Publishing is a subsidiary of Ozone Research Group, Inc.
A Nonprofit Corporation

Library of Congress Catalog Card Number: 98-070936

Publisher's Cataloging-in-Publication
(Provided by Quality Books, Inc.)

Blewett, Stephen E.
What's in the air : natural and man-made air pollution / by Stephen E. Blewett ; with Mary Embree. -- 1st ed.
p. cm.
Includes bibliographical references
ISBN: 0-9640565-2-6

1. Air--Pollution. I. Title.

TD883.B54 1998 363.738'7
QBI98-1071

Printed in the United States of America

Book cover by Armgardt Design

[See back page for ordering information]

11/99

Contents

Foreword

The purpose of this book is to present a wide range of information about air pollution, both the natural causes and that which is created by humans. Every effort has been made to give accurate, scientific data without making moral judgments or taking any political position.

There are differences of opinion among professionals in the field of air pollution as well as among highly respected scientists. Where they are in general agreement on a particular theory, we have presented it as fact. Where they have been in disagreement, we have tried to submit all sides clearly and let our readers decide for themselves which to believe.

The world's leaders also have widely varying opinions as to the causes and effects of polluted air. What we all share, however, no matter what our scientific or political beliefs are, is a common need to breathe this substance we call air that surrounds us and covers the earth.

Before we can determine what or even whether we can do anything about it, we need to know what pollutants are in our air and how they got there.

As Hippocrates said, "There are in fact two things, science and opinion; the former begets knowledge, the latter ignorance."

About the Author

The Air We Breathe is the result of years of research into the available literature on the atmosphere as well as the experience **Stephen E. Blewett** brings to the subject. As both a chemist and meteorologist with degrees from Stanford University and the California Institute of Technology, Blewett is a leading expert in the field of air pollution. He has studied wind patterns, temperature inversions and smog in Los Angeles and other areas throughout the world for over 50 years.

He is the author of numerous scientific papers and articles on meteorology and the causes of smog and other forms of air pollution.

Assisting the author in the preparation of this book was **Mary Embree**, researcher, writer and editor specializing in nonfiction books. Embree has written for television, for various educational video and audio productions, and for periodicals. She is the author of *A Woman's Way: the Stop Smoking Book for Women* (WRS Publishing) and *The Stop-Smoking Diet for Women* and a stop smoking program published and distributed by HealthEdco, Inc.

What's in the Air

Natural and Man-made Air Pollution

CHAPTER ONE

What is Air?

Air is the invisible substance that surrounds us and extends out many miles from the earth as the atmosphere. Our most precious commodity is the air we breathe. The average person inhales an average of 2,850 gallons of air every day. Although we can go without food for weeks or longer and without water for days, we can go without air for only a few minutes. Permanent brain damage or death can occur if the brain is deprived of oxygen for as little as four or five minutes.

Air is a mixture of many gases. One cubic inch of air contains about 300 billion billion molecules. The earth's gravity, however, holds most air molecules within a few miles of the ground where they fly about colliding with each other.

About 99 percent of the air is made up of molecules of two different elements: nitrogen and oxygen. It is around 21 percent oxygen, 78 percent nitrogen and slightly less than one percent argon. The air contains small amounts of carbon dioxide, carbon monoxide, ozone, nitrogen oxides and many other gases such as neon, helium and methane in minute quantities. It also contains water vapor as well as dust and pollen and other kinds of matter.

Humidity is the amount of water vapor in the air. Relative humidity is the amount of moisture in the air at a particular temperature as compared with the maximum amount the air can hold at that temperature. Warm air can hold more moisture than cold air can. Humidity can vary widely from almost none over deserts to 100 percent in heavy fog or rain.

Oxygen, the life-giving element in the air, is one of the most abundant elements on earth. Humans, animals, insects and plants must have oxygen to survive. Even fish respiration is dependent upon the free oxygen dissolved in water. Fires must have oxygen to burn and it is oxygen that supports chemical changes in matter and causes metals to rust. Mammals give off carbon dioxide every time they exhale. Plants take in the carbon dioxide and give back oxygen which helps to create a balance. This exchange from oxygen to carbon dioxide and back to oxygen is known as the oxygen cycle.

About two hundred years ago chemists discarded their "phlogiston" theory in which they believed that burning was the release of a mysterious substance. In 1783, oxygen was defined by the French chemist Antoine Laurent Lavoisier, one of the most honored men in the history of science. Through a series of remarkable experiments and meticulous measurements, he proved that burning, the rusting of metals and the process of breathing consisted of the union of oxygen with other chemicals. His findings were published in 1789 in *Traité élémentaire de chimie* (Elements of Chemistry). Considered one of the most thrilling scientific discoveries ever made, it marked the beginning of modern chemistry.

Air can be compressed or expanded. As air is compressed by its own weight, about half of the bulk of the atmosphere is contained in the bottom layer called the troposphere. The troposphere, which varies in height from the surface of the earth to the stratosphere, is where nearly all clouds and weather conditions occur. There is a lot of space between the molecules, and compressing the air squeezes the molecules closer together. Air will also expand to fill any space that has fewer molecules, as the air's molecules simply spread out and become farther apart.

Air has weight. At sea level and at a temperature of 59 degrees Fahrenheit, a cubic foot of air weighs about .0765 pounds or slightly more than an ounce. One cubic mile of air at sea level weighs about five million tons.

As we cannot put off breathing the surrounding air whether it is polluted or not, we need to learn all we can about what is in the air we are pulling into our lungs. It's important for us to know not only about smog and industrial pollutants but about hazardous indoor air and tobacco smoke. What we breathe is ultimately as important to our health, both short-term and long-term, as what we eat and drink. The quality of the air we breathe is crucial to our very survival. Identifying problems and addressing them promptly can prevent disasters.

Long before we understood what the air was made of, we had air pollution problems. London's coal smoke and fog combined to make a deadly smog for centuries. Even in prehistoric times mankind created smog when smoke from their fires mixed with the fog.

The main elements of air pollution

Hundreds of billions of dollars have been spent trying to clean up the air. The good news is that toxic emissions from coal-burning factories and poisonous substances from industrial effluents have been abated in many areas. Automobiles have been modified to produce less carbon monoxide fumes. As lead is no longer a common ingredient of paints and gasoline, the fumes from these products are safer. The bad news is that there is still air pollution in many areas of the globe that is causing damage to plants, animals and humans. Some of it is of natural origin; some is created by humans.

Smog is a loosely defined term for most kinds of air pollution. However, smog in fact is a coined word meaning the combination of smoke and fog. Not all air pollution is smog. Air pollution can range all the way from being virtually harmless to deadly depending upon its concentration and composition.

What is in our air that can be so damaging to our health? There are so many substances from so many sources that it is virtually impossible to name them all, but most scientists agree that the major elements are as follows:

Sulfur gases

Hydrogen sulfide (H_2S) is a colorless, flammable, water-soluble and poisonous gas that smells like rotten eggs. Because of its offensive smell, it is burned in oil fields to convert it to sulfur dioxide (SO_2).

Sulfur dioxide is a colorless, nonflammable, water-soluble gas which is formed when sulfur or hydrogen sulfide burns. Used mainly in the manufacture of chemicals such as sulfuric acid, it is also used to preserve fruits and vegetables and in bleaching, disinfecting and fumigating.

Most of the major air pollution disasters were caused by industrial fumes that contained sulfur in one form or another. For example, the great smog attacks in London, New York, Donora, Pennsylvania and the Meuse Valley of Belgium all contained many sulfur gases.

Coal gives off sulfur fumes when it burns. In London, New York and Pittsburgh, as well as in other cities around the world, coal was used in home heating and cooking for many years. It was also used in industry.

The sulfur fumes emitted by such extensive use of coal created a great deal of air pollution and even some smog disasters as we will explain in detail later in the book.

Volcanoes emit gases when they erupt. Even when not erupting, cracks in the ground allow gases to vent to the surface through fumaroles. Among these gases are water vapor, carbon dioxide, sulfur dioxide, hydrogen sulfide and hydrogen. Large eruptions force sulfur dioxide gas into the stratosphere. There it combines with water to form an aerosol of sulfurous acid which hastens ozone destruction. It can also lower visibility and entrap particulate material.

Ammonia

Caused by the decay of biological matter, ammonia (NH_3) is a colorless alkaline gas with a pungent odor and acrid taste. In the atmosphere it can unite with sulfur dioxide to form an aerosol.

Hydrocarbons

A hydrocarbon is any one of a large number of compounds (gases, liquids and solids) composed of the elements carbon and hydrogen in various proportions. They were once considered organic compounds because it was thought that they could only be obtained from natural sources. Now, however, many of them are prepared synthetically.

Hydrocarbons vary in chemical activity. It is the hydrocarbons in gasoline that power our cars. They burn in air to yield carbon dioxide, carbon monoxide and water. Natural gas and petroleum are the most important sources of hydrocarbons. When fuel molecules in an engine do not burn or when they only partially burn, it results in hydrocarbon emissions plus other gases and solids.

Polycyclic aromatic hydrocarbons (PAHs) are a group of chemicals formed during the incomplete burning of coal, oil and gas, garbage or other organic substances. They can occur naturally or be man-made. As pure chemicals, PAHs generally exist as colorless, white, or pale yellow-green solids. Found throughout the environment in the air, water and soil, they are considered a health hazard and can enter the body by breathing, eating or drinking substances containing the chemical or from skin contact with it. PAHs have been found at hazardous waste sites and are also in substances such as crude oil, coal, coal tar pitch, creosote and in tar used in roofing and to build roads.

PAHs can be found in the home environment as they are present in tobacco smoke and in smoke from fireplaces. Food which has been grown in contaminated soil or air may contain PAHs as well as meat cooked at high temperatures during grilling or charcoal-broiling.

Most of the exposure to PAHs occur in the workplace as they have been found in coal-tar production plants, coking plants, bitumen and asphalt production plants, coal-gasification sites, smoke houses, aluminum production plants, coal-tarring activities, and municipal trash incinerators. Wherever petroleum products or coal are used or where wood, cellulose, corn or oil are burned, PAHs can be found. The U.S. Department of Health and Human Services considers PAHs carcinogenic.

Carbon monoxide

Carbon monoxide (CO) is a colorless, odorless, poisonous gas produced when carbons or hydrocarbons burn with insufficient air. As a product of incomplete combustion, it occurs when the carbon in the fuel is partially oxidized rather than fully oxidized into carbon dioxide. Because carbon monoxide reduces the flow of oxygen in the bloodstream, it is particularly hazardous to people with heart disease.

Smoke from cigarettes, cigars and pipes contain relatively large amounts of carbon monoxide. Many authorities believe that this may account for the increased incidence of heart attacks and stroke experienced by smokers and those exposed to second-hand tobacco smoke.

During temperature inversions, carbon monoxide from motor vehicles, wood- and coal-burning stoves and industrial emissions may become trapped in the atmosphere, creating a health hazard. This can happen both in high-altitude cities such as Denver and Mexico City as well as low-lying areas such as the Meuse Valley in Belgium as a result of unusual meteorological conditions. CO can be deadly in closed-in spaces and every year there are many cases of defective or unvented gas heaters causing carbon monoxide poisoning and death. This can also happen when people let their cars idle in a closed garage. CO from auto exhausts is not as serious a problem now as it once was because cars have been designed to emit far less carbon monoxide than in the past.

Biomass burning (the burning of vegetation) generates carbon monoxide among other compounds. Emissions are created when land is cleared for agricultural use by burning and from forest fires.

Carbon dioxide

Carbon dioxide (CO_2), a natural constituent of the atmosphere, is a non-toxic gas. It is a product of animal respiration, it is in alcoholic fermentation, and is brought about by the action of yeast. It is what causes the rising of bread dough. In its solid or "frozen" state it is called dry ice or carbon dioxide snow. In motor vehicles, it is a product of perfect combustion. It is also formed in the combustion of carbon or carbonaceous materials such as wood, coal and fuel oil.

Vegetation consumes carbon dioxide and some scientists believe that the ocean dissolves any excess, acting as a regulator. It is found among the gases emitted by volcanoes. Even when a volcano is not erupting gases can vent to the surface through cracks in the ground. As carbon dioxide is heavier than air it tends to collect in depressions and can accumulate in sufficient concentrations to suffocate animals and people.

Nitrogen oxides

Oxides of nitrogen are formed by lightning and by man-made processes involving electrical discharges and high temperature combustion. Due to the high pressure and temperature conditions in an engine, nitrogen and oxygen atoms react to form various nitrogen oxides. They are considered by many to be precursors to the formation of ozone and contribute to the formation of acid rain.

Nitrogen dioxide (NO_2) is also a natural constituent of the atmosphere. In many areas, the earth's soil may be the principal source of NO_2. It is dark chocolate brown at higher temperatures and ranges through reddish brown, orange and yellow to white at lower temperatures. The entire gamut of colors can often be seen in the sky, particularly at sunset. Near the ocean it often appears as a whitish or coppery haze.

Natural oxides of nitrogen are widespread, covering vast areas of the globe. A direct correlation can be found between hazy clouds in the lower level of the atmosphere and smog.

Even in small concentrations, NO_2 has a distinct, acrid odor. It can cause a burning sensation in the nose due to the acid reaction of the moisture on the mucous membrane. It is extremely poisonous and, in large concentrations, can be harmful to the respiratory system. However, it is rare that such concentrations are reached.

Ozone

Ozone is a three-atom molecule of the same oxygen that we breathe. It is colorless, invisible and generally harmless in the concentrations normally found in our atmosphere.

As ozone from the stratosphere gets closer to the ground, much of it is destroyed by contact with smoke, fumes, foliage, and buildings which turn it into oxygen. When the sun goes down and the air cools, the resulting temperature inversion prevents any more ozone from descending to the ground. This is the reason ozone disappears at night. Above the inversion layer, however, in the mountains and foothills, there is often ozone at night.

Only a small amount of ozone blows out to sea. It is new ozone plus other natural contaminants coming in off the ocean that raises the concentration of natural gases. During the day on the west side of many continents the wind blows from the west, over the ocean. At night the wind is often offshore but generally lighter than during the day.

Ozone is ozone wherever you find it. Ozone in the atmosphere at ground level is the same ozone as that in the stratosphere which protects the surface of the earth from damaging radiation. It is sometimes found in greater quantities in rural areas than in cities. Ozone ebbs and flows on a daily and seasonal basis. It is formed in the stratosphere by the ultraviolet rays of the sun and descends into the lower atmosphere.

Large amounts of ozone are created by lightning. Ozone is put into the air in minute quantities by electric motors, sparks and any electrical discharge.

It has been used commercially to destroy tobacco smoke and other offensive smells in hotel and motel rooms.

Some home air-filtration systems utilize ozone. Many water companies use it to purify their water. It is also used by some commercial laundries to clean and purify the wash.

Ozone can be found all over the world, not only in cities but in rural areas far away from any man-made source.

Radon

Radon (Rn) is a chemically inert, radioactive gaseous element present in varying quantities in the atmosphere and in soils around the world and its emissions are considered hazardous to the health. A colorless, odorless and tasteless gas, radon occurs worldwide in the environment as a byproduct of the natural decay of uranium present in the earth. As it can seep into ground water and remain trapped there, well water may become contaminated with radon.

Radon from the soil may enter a home through small spaces and openings such as cracks in concrete, floor drains, wall-floor joints in basements, and the pores in hollow block walls. Concentrations of radon are greatest in areas closest to the ground with the levels decreasing the higher one goes in a structure.

In the home, especially in cold weather when windows are closed and there is little or no air circulation, radon gas and its decay products may accumulate. Radon is less of a problem outdoors because the gas is diffused in the surrounding air.

Prolonged inhaling of radon decay products is associated with an increased risk of lung cancer. Radon gas breaks down into radioactive particles which can become trapped in the lungs. As the particles continue to break down, they release radiation that can damage lung tissue. Interestingly, short-term exposure to high concentrations of radon is considered less of a risk than long-term exposure to a lower level of the gas.

Asbestos

Asbestos is a fibrous mineral which is resistant to acid and fire. It is found in soil and comprising veins in rock and often appears to be a product of metamorphism. As it is strong, durable, and an efficient insulator, it has been used in construction. Tiles for flooring and ceilings, wallboards, exterior shingles for buildings, and insulation for heating and electrical systems are among the products that consist of asbestos either alone or in combination with other materials.

As asbestos has been identified as a carcinogen, the U.S. Environmental Protection Agency (EPA) and the Consumer Product Safety Commission (CPSC) have taken steps to reduce the consumer's exposure to it. Most of the homes constructed in the United States during the past 20 years do not contain materials made with asbestos. By 1996 most asbestos products were banned and the remaining products containing asbestos must now be labeled.

Ingested asbestos fibers lodge in the lungs and, as it is

a very durable material, it remains in tissues. When it becomes concentrated with repeated exposure, it can cause cancer of the lungs and stomach. It is unknown whether any level of exposure to asbestos fibers is safe.

Health risks exist when asbestos-containing materials crumble, flake or deteriorate and minute fibers are released into the air. However, when left undisturbed, there is little likelihood of its causing any damage to the health. Properly installed asbestos-containing house shingles present little or no health risks. When asbestos siding becomes worn or damaged, the fibers can be sealed in by spray painting.

Formaldehyde

A gaseous compound of carbon, hydrogen and oxygen, formaldehyde is colorless and has a strong suffocating odor. It is used in the preparation of certain dyes and in the production of synthetic substances such as bakelite and other plastic and synthetic resins. It is also used as a corrosion inhibitor in oil wells.

Formaldehyde is prepared commercially by passing methyl alcohol vapor mixed with air over hot copper or other substance that acts as a catalyst, causing partial oxidation of the alcohol.

Many construction materials and consumer products contain formaldehyde-based glues, resins, preservatives and bonding agents. Until the early 1980s it was an ingredient in foam used for home insulation. It has been used in making permanent-press fabrics and as a preservative in some paints and coating products.

In the home, sources of formaldehyde include tobacco smoke, unvented fuel burning appliances such as gas and kerosene stoves and heaters and household products. Perhaps the most significant sources are from the adhesives used to bond pressed wood building materials and plywood for furniture, interior cabinets and exterior construction.

Formaldehyde has been shown to cause cancer in animals. It can trigger asthma attacks and other breathing difficulties in humans as well as cause skin rashes and burning sensations in the eyes, throat and nasal passages.

In some cases, increasing the ventilation and improving the circulation of outside air through the home may be sufficient to reduce formaldehyde to acceptable levels. In others, removing the formaldehyde-bearing material may be necessary.

Lead

Lead is a heavy metallic element found in rocks and soil throughout the world. The color of light silver when freshly cut, lead darkens when exposed to the air. It is relatively soft and malleable and a poor conductor of electricity and heat. Because of its very high density, lead is used as a protective shield against X-rays and radiation from nuclear reactors.

Since ancient times it has been known that lead is toxic. Airborne particles can enter the body through breathing or swallowing lead dust and accumulate in soft tissue. High concentrations can cause permanent damage to the central nervous system, brain, kidneys and red blood cells, and even death. It is believed that even low levels may contribute to high blood pressure. Because lead is more easily absorbed into growing bodies and their tissues are more sensitive to the damage, fetuses, infants, and children are more vulnerable than adults. It may take a relatively low concentration of lead to damage a child.

Until gasoline was reformulated without lead, automotive exhaust was the leading source of airborne lead dust. Many paints also contained lead until recently. It can be present in drinking water, in soil and in house dust.

Pollen and other airborne substances

For thousands of years or more, people have suffered from natural airborne pollutants such as pollen. These airborne irritants, called allergens, trigger the release of histamine, a body chemical. Reactions range from inflammation of the nasal passages, sinuses and eyelids to hypersensitive reactions in the lungs and chest. Allergic reactions can be caused by ragweed, dust, molds and pollen from trees, grass and flowers.

Not only pollen, but any airborne substance derived from a living organism can cause irritation of the eyes, nose and throat, difficulty in breathing and other reactions. Among them are hair, feathers, animal dander, and tiny mites that infest house dust.

Dust

Our normal environment contains dust-laden air whose particles vary from time to time in number and chemical and physical properties. As the soil contains a wide variety of natural pollutants as well as those made by humans, it can be hazardous to the health.

An increase in Valley Fever was linked to dust clouds kicked up by the Northridge (Southern California) earthquake in 1994. The dust carried fungus spores that caused a deadly outbreak of the flu-like illness. Between January 24 and March 15, 1994, there were 203 cases of Valley Fever. There had been fewer than 60 cases during all of the previous year. The symptoms can vary. Some people experienced only mild symptoms while others felt fatigued or had respiratory problems similar to the flu; three residents died of the disease.

Tornadoes or even moving large amounts of dirt during major construction projects may create health problems in areas where the fungal spore or other toxic materials are present in the soil.

In many areas leaf blowers have been outlawed, not only because of their noise pollution but because of the dust they stir up.

Particulate matter

In the early 1990s the Southern California Air Quality Management District began to focus on particulate matter, a mystifying blend of man-made and natural substances. Regulators named it PM 10 because it is smaller than 10 microns, one-fifth the width of a human hair.

This hazardous component of our air consists of several things including many if not all of the pollutants mentioned above. By the mid 1990s, regulators had included PM 2.5 which they considered even more of a problem because the particles were smaller and could work their way deep into the lungs, potentially causing more damage than PM 10.

Fine, black particles comprised of soot from trucks and buses, ammonia from dairy farms, sulfur from factories, dust and dirt from streets and construction sites, ash from fires and even sea salts have clogged the filters in Rubidoux, a monitoring station on the outskirts of Riverside in Southern California. This mass of airborne particles often reached three to four times the health standard set by state authorities.

The American Lung Association noted that particulates have taken a back seat to ozone for years and regulators will have to play "catch-up" in developing control programs to deal with them.

Particulates and ozone remain among the most poorly understood of all the major air pollutants.

Natural Sources of Air Pollution

Since ancient times there have been deaths caused by natural gas emitted from fissures in rocks such as those about 100 miles west of Babylon in the town of Hit where natural asphalt was mined for a profitable building industry. Egyptian King Tukulti visited the town in 900 B.C. and tells of camping "where the voice of the gods issueth from the ulmeta rocks."

When asphalt bitumen oozed into rivers and wells where the ancients got their water, they associated it with the underworld. They believed it was a symbol of powerful evil spirits who were rising from deep inside the earth to torment mankind. The Bible speaks of "the lake of pitch which is hell."

For as long as there has been an earth, there have been

volcanic eruptions that have tossed gases, ash, dust, molten lava and solid fragments into the air. Fluorine poisoning caused by volcanic eruptions thousands of years ago has been found in human bones in Iceland.

Volcanoes

In 1815, Tambora in Indonesia created an ash cloud that was blamed for a cold summer in Europe and North America the following year. Consequina in Nicaragua hurled two and a half cubic miles of ash into the air, darkening all of Central America for two or three days. Volcanic material from Krakatoa, near Java, was sent 17 miles into the atmosphere and was carried around the entire globe. For several years afterward, the volcanic particles in the air created unusually beautiful sunsets.

In 1991, Pinatubo in the Philippines erupted after being dormant for 600 years. It was reported at the time that it temporarily cooled the earth's climate by half of a degree on average and contributed to ozone depletion. In fact, in this century, according to the experts, several eruptions have been responsible for lowering the world's average temperature by as much as one degree centigrade for a number of years.

Popocatepetl, in the vicinity of Mexico City, erupted in December of 1994 and continues to erupt, spewing out ash over a wide area. High emissions of sulfur dioxide gas from the volcano are still being reported. They range from 1,000 to 12,000 metric tons a day. Only Mt. Etna in Sicily has emitted that much sulfur dioxide gas during non-eruptive periods.

On July 18, 1995, the Soufriere Hills volcano on the tiny eastern Caribbean island of Montserrat roared to life after being dormant for nearly four hundred years. In late June of 1997, fiery eruptions of hot rock, gases and ash claimed 19 lives, its first casualties. In August, dramatic pyroclastic flows of rock and ash swept into Plymouth, the abandoned capital, setting fire to its historical old buildings and filling the streets with volcanic debris and overwhelming the port. Continuing eruptions send ash clouds over 10,000 feet into the air.

In Guatamala the Pacaya volcano erupted in May of 1998 mixing volcanic ash with the haze that is normal in the region during the spring months. The volcano, which is 12 miles south of Guatamala City, disgorged 1,000-foot columns of smoke into the air, limiting visibility and forcing aircraft to use instruments for landings. The smoke, which traveled as far as Mexico and the Southern United States, was particularly bad in Honduras where it closed the country's two main airports for over 24 hours.

Volcanic gases

Ash clouds and gas emissions from the more violent eruptions are sent high into the stratosphere where they circle the globe. Extremely large quantities of sulfur are emitted by some volcanoes. They form natural aerosols that contribute to the depletion of the ozone layer. It is ozone that protects the earth from ultraviolet radiation. Man-made effluents are dwarfed by these eruptions.

Volcanoes don't have to be erupting to emit gases. Cracks in the ground allow volcanic gases consisting of water vapor, carbon dioxide, sulfur dioxide, hydrogen sulfide, hydrogen and other gases to vent to the surface through holes called fumaroles. Sulfur dioxide gas can mix with water droplets in the atmosphere and fall as acid rain. As carbon dioxide is heavier than air it can collect in depressions in lethal concentrations where it can cause suffocation.

Pliny the Elder, the ancient Roman naturalist and writer who lived in A.D. 23-79 is believed to have been suffocated by volcanic fumes when Vesuvius destroyed the city of Pompeii with ash falls. People have been found still sitting at their dinner table, seemingly uninjured from the volcanic eruption, but who had died on the spot from carbon dioxide suffocation.

Lightning and wildfires

There are many natural events that can create toxicity. Lightning makes tons of nitrogen dioxide and ozone.

Lightning also causes forest fires. In fact, far more fires are caused by lightning than by humans. In mid-1998 the worst outbreak of wildfires in 50 years occurred in Florida. The smoke from the fires was so thick that fire-fighters couldn't tell where or how many fires there were. Hundreds of residents were forced from their homes as smoke blanketed the region from Orlando to Jacksonville burning eyes and lungs and creating a health hazard.

Extreme weather conditions

Contributing to the problem was the unusual weather system called El Niño which causes torrential rains in some areas and droughts in others. The notorious El Niño of 1997-1998 made half of the world a tinderbox, according to an article in the June 22, 1998 issue of Time, a monthly magazine.

By June of 1998, uncontrollable fires had roared through 7,700 square miles of Indonesia. Fires had devoured 1,460 square miles in Canada and consumed 2,150 square miles in Central America. In Mexico, more than 11,000 fires burned 1,500 square miles during the worst drought the region had experienced in 70 years. Forest fires burned around Mt. Kilimanjaro in Tanzania and in Kenya, Rwanda, Congo and Senegal. In Russia, an unusually large outbreak of fires threatened Siberian tiger habitats. Brazil experienced catastrophic wildfires which engulfed more than 20,000 square miles, much of it in the already endangered Amazon rain forest.

Although most of the fires were caused by natural conditions, human activity contributed greatly to the destruction. Logging practices which leave combustible material on the forest floor create incendiary conditions. As the forest canopy is thinned out, sunlight dries out the vegetation and virgin rain forests that are usually humid have been burning, often for the first time in history.

Roads that have been carved through tropical forests provide access to loggers, farmers and ranchers, all of whom use fire to clear land. In recent years, more and more of these fires have gotten out of control. (See Chapter 3, "Man-made Pollution.")

The ocean, land bogs and marshes

A relatively unknown source of air pollution is our oceans. According to the Geophysical Institute Quarterly, northern oceans may be releasing large amounts of natural gases into the atmosphere. "Sulfur particles can change the properties of clouds, and affect global change," according to an article in the Quarterly.

Assistant Professor of Chemistry Richard Benner of the Geophysical Institute has designed an instrument that can detect sulfur in air and in liquid. Although he developed the instrument solely for atmospheric studies, it has been sold commercially to breweries and wineries where it is used to measure sulfur compounds that create distinctive tastes in beverages. His sulfur-detecting instrument is challenging theories about the source of sulfur in the atmosphere. Most experts believe that about half of the earth's airborne sulfur is derived from man-made sources and half from natural sources. Benner's research, however, indicates that up to two-thirds of the sulfur in the earth's atmosphere comes from natural sources such as land bogs, and a major portion of it comes from the ocean. Turbulent waters and air currents appear to be the main mechanisms for mixing sulfur in the air.

Weather blamed for the extinction of dinosaurs

Some of the theories regarding the extinction of the dinosaurs have to do with natural air pollution. One theory is that an asteroid from outer space struck the earth during the Mesozoic Era. The impact caused an enormous cloud of dust to circle the earth striking at the food chain by killing many plants.

Other geologists believe that there were many climate-changing volcanic eruptions over a longer period of time. They theorize that as the volcanic clouds blocked out the sun, the vegetation died out and the dinosaurs died of starvation. The air may also have carried toxic fumes and suffocating ashes which poisoned their lungs.

Directly or indirectly, it was probably some form of air pollution that killed the dinosaurs.

Natural smog has been observed about halfway between New Zealand and Los Angeles over the Pacific

Ocean at altitudes of over 20,000 feet. Because of the remoteness of the location, this could not be from any man-made source such as cars or industry. Some seamen have reported that the ventilation system on surface ships has drawn in air and discharged it throughout the ship and the smell is much like Los Angeles smog.

Clara Jodwalis, a doctoral graduate student in atmospheric chemistry, has been studying airborne sulfur. The intensive process of collecting samples above the Atlantic Ocean near the Azores took nearly two years to complete. Her results also indicated that there was significantly more sulfur in the atmosphere from natural sources than had been previously believed and that the northern oceans appear to be the major sources.

There is evidence of hay fever and other diseases in early humans caused by plant pollen and gases emitted by marshes. Ancient literature refers to "miasmas" which could have been noxious fumes from rotting organic matter. The air near the floor of a dense forest covered with decaying leaves can contain three times the normal amount of carbon dioxide.

In closed spaces such as crowded rooms, elevators and subways, the concentration of carbon dioxide can be a great deal higher than normal. The concentration can be lethal in caves and at the bottom of wells. In Calcutta, on a summer night in 1756, prisoners were put into the "Black Hole," a room that was 18 x 14 x 6 feet in size and contained only one small window. Of the 146 prisoners left in the small room that night, only 20 were still alive by morning.

Sulfur particles can be harmful when released into the environment. They can cause a haze of air pollution dense enough to hinder visibility. According to Benner, "When you are standing on the rim of the Grand Canyon, or looking across the Alaska Range on a clear day, but you can't see a thing through the haze, you are looking at particles mostly made up of sulfur."

Ozone (O_3) is also a natural constituent of the air. It is formed in the stratosphere from oxygen (O_2) by the sun. A portion of this ozone subsides (descends) to the surface of the earth in varying quantities. As the sun heats the ground, it overturns the air, bringing down the ozone hovering above, often creating a severe smog problem.

The type of killer smogs that occurred in London, England and the Muese Valley in Belgium in the 20th century were not the same as those experienced in cities such as Los Angeles, Santiago and Rio de Janeiro. In coastal areas like these there can be eye-stinging, acrid air without any industrial sources. In late spring, in summer and early fall ozone descends from aloft, mixes with oxides of nitrogen and other natural impurities, and can be held close to the ground by an inversion layer and motionless air. The resulting smog can be very distressful.

The Valley of Smokes

In 1542, when Juan Rodriguez Cabrillo sailed into what is now San Pedro Bay in Southern California, he named it *Bahia de los Fumos.* Long before that, early Native Americans called the region the Valley of Smokes. To escape the naturally occurring smog, they moved to the mountains during the warmer months.

In 1868, the editor of "Alta," a leading California newspaper wrote:

> "It is now about six days that we have in this and surrounding country been spectators of an unusual atmospheric phenomenon, which from its peculiarity, has given occasion to many manifold surmises, conjectures, speculations, and rumors..."

What the editor saw, most likely, was what is now called smog.

Man-made Pollution

Air pollution created by humans probably began when they first discovered how to manage fire. They found that they could capture flames by putting the end of a stick into an existing fire and holding it there until it burst into flame. They could then wave the firestick about to scare away wild animals. It was probably soon after that use for fire was found that they learned to light a pile of dry wood with the burning stick and use the resulting fire for heating and cooking. The harnessed fire provided humans with light to see by on dark nights Its mystical properties also enhanced their primitive religious rituals.

At first early humans used fires that had been started by lightning. They carried it from place to place and learned to maintain it for long periods of time. Eventually they learned how to start fires themselves by rubbing sticks or stones together.

"Slash and burn" farming

For centuries and possibly even longer, farmers have used a process called "slash and burn" in which they clear land for planting by setting brushland and forests on fire. During droughts or unusually hot weather, slash and burn farming technique can set off unintended fires even in tropical forests and jungles that can run out of control.

In early 1998, partially due to the El Niño weather phenomenon, fires set in Mexico created raging forest blazes that reached catastrophic proportions. The fires were blamed on farmers clearing the land and on drug traffickers who set fires as a diversion technique. Thousands of fires burned out of control for months, triggering the U.S. and Mexico to jointly make a disaster declaration on May 15.

The experts called the fires an environmental catastrophe. The U.S. Agency for International Development (USAID) considered it "the most serious of its kind in the world ... and the most difficult to fight." Authorities in

Mexico, fearful that they were losing an ecosystem that had taken thousands of years to form, considered it Mexico's tragedy of the century.

Clouds of smoke from the from the fires in Mexico and Central America drifted north and blanketed Texas, reducing visibility to less than three miles in some areas. State officials issued health warnings, advising residents, especially children and the elderly, to stay indoors as much as possible and to avoid exercising outdoors. Smoke from the fires reportedly drifted into Florida and as far north as Illinois and Wisconsin.

Human-created air pollution is as old as the history of fuel. In 361 B.C., the Greek philosopher Theophrastus noted that the "fossil substance called 'coals' burn for a long time, but the smell is troublesome and disagreeable." The poet Horace lamented in 65 B.C. that the shrines of Rome were being blackened by smoke.

Medieval smog

For many centuries London has been plagued with air pollution caused by human activity. When the present metropolis was still only a walled, medieval town it had already gotten a reputation for having "bad air." It began during the Plantagenet times in the 13th century with the introduction of bituminous coal. As wood had been used both for building and for fuel for centuries, the great forests had been greatly depleted by the late twelfth century. The price of wood increased drastically and an additional source of fuel was needed.

A convenient substitute proved to be "sea coal," so named because it had been found in abundance by the northeast seacoast and the Firth of Forth. Brewers, dyers and artisans began burning the new sea coal in their kilns and furnaces. By 1228, the practice was so common that a suburban byway just outside the town was named Sacoles (or Sea Coals) Lane.

Coal did not come into common use as domestic fuel for another century or two. But as wood became scarcer and more expensive, the poorer classes began to burn coal in their homes.

English and Scottish cities have long suffered from smog. It was reported that in 1909 there were 1,063 deaths from "smoke-fog" in Glasgow and Edinburgh. London, however, has been hit the hardest because of the additional problems with temperature inversions which keep pollutants close to the ground for longer periods of time.

During the dark, cold days between November 12 to 25 in 1922, heavy smog was experienced throughout many parts of Great Britain, and in London the death rate from respiratory diseases more than doubled.

In 1948, as a result of unrelenting smog in Donora, Pennsylvania, more than 5,000 persons fell ill and 20 died. In London, England, smog accounted for the deaths of more than 4,000 in 1952 and 106 in 1962. These deadly smogs, however, were of a very different kind than the smog in regions like Southern California.

The killer smog attacks throughout the world, such as those that occurred in Donora and London usually come about because of stagnating air, a very low and intense temperature inversion, smoke, and chemicals spewed out by industry and the public.

Coal and wood burned in home furnaces and fireplaces also can contribute to air pollution problems. In Donora and London the lethal attacks had little or nothing to do with natural ozone or nitrogen dioxide as they were prevented from reaching the surface in any concentration by the same inversion that held the smoke and fumes below.

Generally, smog is of two types: The London/Donora type where the air is polluted by compounds produced by burning coal and industrial effluents and the Los Angeles type where a mixture of oxides of nitrogen and ozone descend from aloft and also move inland from the ocean.

The Muese Valley in Belgium

Across the English Channel, on December 1, 1930, a thick and persistent fog engulfed Belgium. Temperatures were low, the air still and above the Muese River a strong thermal inversion trapped the smog between the steep, narrow sides of the river valley. It affected a fifteen mile stretch of the Muese valley. Small villages and farms dotted the area. But it also contained steel mills, glass factories, lime furnaces, power stations and plants which produced zinc, sulfuric acid and chemical fertilizer, their stacks pumping a myriad of waste products into the atmosphere.

By the third of December, thousands were feeling great distress. They started vomiting and gasping for breath. Before the smog finally dispersed on December 6, several hundred people had become seriously ill and sixty had died.

Around the world the Meuse valley disaster created headlines. How could it have been an ordinary smog that sickened and killed so many, so quickly and so violently? There were rumors that mysterious airplanes swooped down over the valley and dropped some sort of poison gas. Some thought it must have been caused by a factory accident where lethal fumes leaked from a broken pipe. One of Britain's most eminent scientists, Dr. J.B.S. Haldane of Cambridge even suggested that it might have been an outbreak of the bubonic plague.

An investigation was immediately launched by the Belgium government. Within a short time they dispelled the rumors of a mysterious airplane, a factory accident or the plague. It was nothing other than an acute air pollution disaster. No single pollutant was identified as the culprit. More than thirty impurities were present in the smog, most of them poisonous and injurious to health. Although the fatal irritant was never determined for certain, authorities believed that it had been sulfur dioxide gas and sulfur trioxide aerosol emitted by the factories.

Perhaps it was because of the Meuse valley incident that Londoners were not as concerned about their own dirty air as they should have been. Their smogs, they believed, were different. Although London fogs did contain sulfur dioxide and industrial pollutants, they were principally characterized by soot, carbon particles and smoke from raw coal.

The kind of the smog the Meuse valley experienced was not likely to occur in London. The geography was different. The Thames was more open and broader than the Meuse.

The gathering clouds of catastrophe

During the early war years (World War II) the Ministry of Home Security evolved an ingenious scheme to conceal important targets in and around London from Nazi bombers. Some of the factories were ordered to produce as much smoke as possible from their chimneys to deliberately create additional smog. By 1943 a fuel shortage put an end to the practice and the Ministry then asked for the maximum amount of work for the minimum amount of coal.

After the war, most concerns about air pollution were put on a back burner as the country began rebuilding. Year by year, London's air became more heavily contaminated. Then, on Thursday, December 4, 1952, an extensive high-pressure weather system spread slowly in a southeasterly direction across most of the British Isles. With it, the system brought light variable winds and low ground temperatures.

By dusk, the center of the system lay about two hundred miles northwest of London. The winds died down in the valley of the Thames during the night and the area experienced a marked temperature inversion. Air near the ground grew colder and became trapped beneath a lid of warmer air above. Unable to rise and with no wind to disperse it, the cold air remained totally inert.

London's air pollution disaster

Late Thursday evening a dense smog began to form and by next morning as the city awoke, tons of smoke from millions of domestic chimneys were discharged into the foggy, cold and motionless air. To this was added tons of coal smoke and sulfur oxides from huge power stations. Factories and industrial plants contributed their pollutants as well.

Within a few short hours the air had become massively contaminated by a lethal mix of smoke, soot, carbon particles and gaseous wastes. The air turned yellow, then amber and then black.

By early evening, the first of the city's inhabitants began to die. For four days and four nights, in what would become the worst air pollution disaster in history (up to that time) the great killer smog held London in its grip.

Nearly one out of every two thousand people in the city died during those four days or perished during the next two weeks. Many Londoners became seriously ill, recovered temporarily, then died months or even years later from the lingering effects of the scourge. Over four thousand people died during the smog attack and hundreds, perhaps thousands more, died later.

In a January 1954 report, the Ministry of Health stated:

> "Not until the death certificates were assembled and analyzed did the extent of the excessive mortality become apparent.... It must in truth be a supreme example of the way in which a Metropolis of eight and a quarter million people can experience a disaster of this size without being conscious all the while of its occurrence."

The Donora, Pennsylvania smog attack

In 1948, toward the end of October, a severe, low temperature inversion occurred over a wide area of the northeastern United States. Along the Monongahela River about thirty miles south of Pittsburgh a heavy fog formed. It was particularly dense in Donora, an industrialized community which had several industrial plants including a large steel mill and a zinc reduction plant and in Webster, a small community on the other side of the river. Like the Meuse, the Monongahela River valley was narrow and lay between steep sloping cliffs. Observers soon noticed that the smoke from the factories and railroad engines began spilling to the ground and were trapped there by the cold, motionless air.

By Friday, October 29 the "Big Smog" had built up into a dirty gray mass. That evening many of the townspeople turned out for the annual Halloween parade unaware that this smog was worse than anything they had ever seen before.

The town's doctors were already being swamped with calls, some of them being from people who had collapsed at the parade. Their complaints had started with a dry cough, headache, sore throat and watery eyes and progressed to chest pains and extreme difficulty in breathing.

By Saturday, forty-two percent of the town's population of 14,000 had been affected. Ten percent were severely affected. Three men became seriously ill on boats passing by on the river.

Before the smog lifted, it was reported that pregnant animals began having miscarriages. According to Public Health Service's Bulletin No. 306, "Air Pollution in Donora, Pennsylvania: Epidemiology of the Unusual Smog Episode of October 1948," a large number of pets and farm animals became ill and many of them died. In all, according to newspaper reports at the time, nearly 1,000 animals were killed by the fumes.

It was four days before a cleansing rain dispelled the smog. During those four days, however, nearly half of the inhabitants had become ill and 20 had died. The United States Public Health Service conducted a thorough investigation and found that Donora, Pennsylvania had experienced an acute air pollution disaster similar to the Meuse valley disaster of 1930. The poisonous air had swiftly produced severe illness and death. Most of those killed had been middle-aged cigarette smokers or elderly people who suffered from a preexisting heart or lung condition.

But all were affected—even the young and healthy.

The Donora tragedy showed with scientific certainty that human beings could be killed or made seriously ill by air pollution. Blame, however, could not be placed on a single pollutant, although many sulfur gases were present. The particulates included a variety of grits, fly ash and soots.

Dr. Clarence A. Mills of the University of Cincinnati wrote an article in the periodical, Hygeia, after visiting the area around Donora six days after the killer smog attack. Dr. Mills was head of the Department of Experimental Medicine in the University's College of Medicine and an expert on harmful effects of air pollution and the effect of climatic changes on civilizations.

As he descended from the upland into the Monongahela River valley he reported entering a "sea of foul, irritating smog." He turned on the car lights and "crept along at a snail's pace," unable to see more than a few feet ahead of him.

As he approached Donora, Mills commented that he noticed:

> "... a terrible, sickening devastation wrought by the acid fumes from the large zinc smelter along

> the river at the lower end of town....
>
> "All vegetation was dead along a wide sweep of hills back from the eastern bank of the river, and devastating erosion had gutted the sloping wold. On the Donora side, what had once been the city's beautiful hillside cemetery was now a gruesome and grizzly spectacle. Trees and shrubs dead, grass gone ... here was a most convincing picture of what three decades of poisoned air could accomplish."

Dr. Mills stated that the Donora smog barely missed being a major catastrophe. In a December 25, 1948 New York Times article he wrote:

> "A slightly higher poison concentration in the air or a few hours longer time and the whole community might have been left almost devoid of life."

Poza Rica, Mexico

A lethal incident occurred in another part of the world on November 24, 1950. In Poza Rica, Mexico, 22 people died and 320 were hospitalized after an industrial accident at a petroleum plant where sulfur was being recovered from natural gas. When efforts were made to increase the rate of flow through the sulfur-removal unit, a surge in the gas flow resulted in the release of large amounts of unburned hydrogen sulfide into the air. Those in close proximity to the effluent stack were poisoned by the fumes. Hydrogen sulfide, one of the most toxic of the sulfur gases, kills fast.

Again, weather played a part. In the area, not far from Mexico City, there was a pronounced ground temperature inversion with a high concentration of haze. There was very little wind movement to dispel the toxic emissions.

The air pollution disasters of London, England; the Meuse Valley in Belgium; Donora, Pennsylvania; and Poza Rica, Mexico were all products of the combination of man-made pollutants, geographical location and meteorological conditions.

In 1951, a report prepared by the Stanford Research Institute blamed the public for 60 percent of the smog in Los Angeles. The other 40 percent of blame was laid to industry. The Western Oil and Gas Association paid for the research for the SRI report. Almost two-thirds of the 2,280 tons of chemicals entering the air each day were organic in nature, the report concluded, and the public accounted for 76 percent of the organic part. Automobiles and buses, and the heating of homes and offices were large contributors to the pollution. However, the biggest single source, the report stated, was the 550 tons of organic matter resulting from the burning of 4,000 tons of trash in backyard incinerators.

Waste disposal

People were ordered to shut down their backyard incinerators and put their trash out to be picked up. A few companies reaped huge profits by getting the contract to haul the stuff away to landfills.

Some waste-disposal managers were opposed to the banning of backyard incinerators. Disposing of so much waste through burying it in landfills, they knew, would create other kinds of environmental problems. Most of the waste materials do not break down and landfills can fill up rather quickly. Toxic substances in the trash can seep into groundwater, polluting it.

Their prediction has proven to be true. Southern Californians are now having to deal with the problem of trying to clean up the polluted groundwater, a potentially greater health hazard than the present air pollution.

It has become increasingly difficult to find appropriate waste disposal sites. No one wants a smelly, potentially toxic landfill to be located in their neighborhood. Environmental groups don't want them to pollute pristine wilderness areas. In the meantime, the present landfills are running out of space.

There were some scientists during the 1950s who went into the field and measured ozone concentrations at various locations and elevations. They concluded that most of the ozone in the lower atmosphere was almost entirely of natural origin. The regulators, however, accepted the opinion that chemical reactions could lead to the formation of ozone in the free surface air and began to regulate the emissions of certain substances. The type of smog such as that which exists in Southern California has not directly been blamed for any killer smogs even though the Los Angeles area is said to have the worst air quality in the nation.

Many of the constituents of air pollution are man-made and can be regulated and controlled. Pittsburgh, Pennsylvania, once called the "smoky city" was a dreary town painted in shades of gray due to the thick, acrid smog caused by burning coal and factory emissions.

Black soot enveloped the buildings and invaded the lungs. In the late 1940s, however, an urban redevelopment program was begun which included strict measures for smoke control and today the buildings are no longer grimy and dingy. They have cleaned up their air and Pittsburgh has become a jewel of a city, sparkling with color.

The Pittsburgh smog was man-made and could be controlled. The kind of smog that appears in Los Angeles, California, in Santiago, Chile, in Rio de Janeiro, Brazil and other coastal cities is largely a phenomenon of nature. These areas get their smog from a mixture of oxides of nitrogen and ozone which move inland from the ocean and/or descend from above the earth to ground level. This naturally occurring smog usually becomes trapped between a seacoast and nearby mountains and is worsened by a sea air temperature inversion.

Smog throughout the world

Acute air pollution episodes associated with temperature inversions have been experienced in many large cities such as Tokyo, Yokohama, Paris, Moscow, Leningrad, Denver, San Francisco, Buenos Aires, Vienna, Mexico City, Calcutta, Milan, Sydney and Osaka. New York,

Connecticut and New Jersey had severe smog for three consecutive days and nights during the Thanksgiving weekend in 1966. Five times the normal level of sulfur dioxide was reported as well as smoke and deadly carbon monoxide which were recorded in amounts well above the danger level.

In 1953, 240 people were said to have died in New York City and a decade later more than 400 deaths were reported as a result of a five-day smog.

Although air pollution experts have known for decades, if not centuries, that smoke from burning wood and coal is a health hazard, toxic emissions from these sources continue to this day in the United States and elsewhere in the world. In Missouri 80% of America's barbecue fuel, charcoal, is produced in 374 giant kilns. The billowing smoke blankets the beautiful landscape of the Ozark Mountains, depositing an ugly residue of black, gritty soot. The open-air burning is blamed for most of the air pollution in the Ozarks. Somehow they have escaped regulation for years but it finally caught up with them in the fall of 1996 when the kiln operators agreed to install afterburners.

The kilns of one of the producers was found to be

emitting pollutants on the town of Moody that vastly exceeded federal air quality standards. According to a local EPA officer, measurements were among the highest he had ever seen. They surpassed air particle pollution readings taken on a very smoggy day in Los Angeles.

In many parts of the world the environment is in jeopardy due to chemical spills and emissions. In the 1960s, Dr. Walter Orr Roberts, then director of America's National Center for Atmospheric Research stated that there was an "imminent likelihood" of an air pollution disaster somewhere in the world which could take as many as ten thousand lives.

His warning proved to be prophetic for, decades later, industrial poisons did kill several thousand people.

Major industrial accidents

What some believe to be the world's worst industrial accident in history occurred in Bhopal, India on December 3, 1984. Due to a defective insecticide storage tank at a Union Carbide plant, tons of lethal gas were blown over the nearby residents during the early morning hours while most were still sleeping. The release of methyl isocyanate into the air during a temperature inversion caused at least

3300 deaths and over 20,000 illnesses. The injuries included permanent damage to eyes and lungs. Many of those who became ill later died.

The following October, a government study found that survivors of the disaster had suffered chromosomal changes and damage to their immune systems. It was estimated that thousands of others will suffer debilitating illness for the rest of their lives. The final tally, should it ever be made, could very well be much higher than the original estimates of fatalities.

Lawsuits claiming more than $200 billion in damages were filed against Union Carbide in the United States as well as in Indian courts. Although the company offered several million dollars in immediate aid to the victims, the offer was rejected by the Indian government on the advice of their lawyers who were going after a much bigger settlement.

After years of litigation, in February 1989 the United States owners of the Union Carbide plant agreed to pay the Indian government $470 million which had been ordered by the Indian Supreme Court in full and final settlement for the disaster. The Indian government, in return, agreed to drop criminal charges against the company and its former chairman. There were some 500,000 claimants to the money. There have been varying reports as to how much of this money ever actually reached the victims.

The world's worst nuclear-reactor disaster was on April 26, 1986 when the No. 4 reactor exploded at the Soviet Union's Chernobyl nuclear power plant. It was while an experiment was being conducted with its emergency water-cooling system turned off. Due to a series of miscalculations, the reactor's heat began to rise rapidly. There was a neutron build-up in one area of the core and because of the pressure, the top of the reactor was literally blown off. A second explosion spewed radioactive waste a mile into the air. The explosions caused 30 fires including one near the No. 3 reactor which was still operating. Thirty-one people were reported killed immediately or shortly afterward with another 500 hospitalized. A radioactive cloud caused by the catastrophic meltdown spread out over most of Europe, contaminating crops and livestock. Radioactivity from the disaster was detected as far away as Asia and North America.

Over 650,000 persons involved in the cleanup were exposed to large amounts of radiation. Some 10 million who live in the most affected regions nearest the plant were also exposed to hazardous levels of radiation. Because the cancer-inducing effects of radioactive fallout cannot be detected for many years, the full impact of the accident has yet to be determined.

The above disasters are examples of unusual and extreme kinds of air pollution which are not usually mentioned in discussions about air pollution. What most people think of as air pollution is what regulators report as smog. Yet, the toxins put into the air can be far more deadly than "ordinary" smog.

The London and Donora type of smog contained coal and industrial emissions. The Los Angeles type is mainly a mixture of natural ozone and oxides of nitrogen, although there may also be man-made pollutants in the mix. However, meteorological conditions, such as temperature inversions, play a major part in both types of smog.

It is important, however, to point up a parallel between London smog and Los Angeles smog. In both cases, the governments were reluctant at first to blame industrial plants. To regulate or relocate the factories that were spewing out huge amounts of smoke and toxic fumes presented too many problems. They were a major source of revenue for the local governments and the industries had large amounts of money to spend to maintain their status quo. In both cities, the smog was first laid to the public.

In London, it was blamed initially on people who burned coal in their homes to cook with and keep warm. Although that certainly contributed to the air pollution, the major sources of emissions were the industries that not only burned coal but discharged tons of toxic substances of many kinds, which affected everything and everyone within miles.

Too often, air pollution studies have been conducted by the very companies that are doing the polluting and they have a vested interest in not finding themselves culpable.

Some scientists believe that human beings are on a collision course with the natural world. Whether or not we agree on the amount of destruction humans have caused, it may be imperative to our very survival that we take responsibility for preserving and conserving what we can.

Automobile Pollution

Automobiles, buses, trucks and other gas-burning vehicles have been blamed for much of the air pollution, especially in densely populated cities. Older, smoking cars still put out foul odors as well as carbon monoxide fumes which can be lethal in closed-in spaces. However, auto manufacturers have now designed catalytic converters and emission control devices which have reduced carbon monoxide and other exhaust gases greatly in vehicles manufactured in recent years.

Implemented first in California, devices to control volatile organic compounds (VOCs) from tailpipe emissions are now required by law in other parts of the United States. Usually installed in new cars at the factory by auto manufacturers, catalytic converters have cut down considerably on carbon monoxide, the most toxic of auto emissions. Carbon monoxide is clear and colorless and may be in the smoke that some cars and trucks emit. A smoking car may not necessarily cause smog. Although there have been numerous studies which indicate that not only do automobiles not contribute to the formation of ozone, they actually destroy it, the pervasive belief in the United States is that cars cause ozone.

Even though American car makers had been forced by foreign competition and American car buyers to make smaller, more fuel-efficient cars, the smog devices added to the fuel consumption.

Vapor recovery devices have also been installed on gasoline station pump nozzles to control gas vapors during refueling. This was done to protect the public and service station attendants from toxic fumes as well as to protect the atmosphere from the increased ozone which most regulators believe the vapors create.

In the United States leaded gas is no longer available. As there are still some cars on the road, as well as lawn mowers and other small motors, that require leaded gas, an additive is sold as a lead substitute so that unleaded gas may be used. Although lead increases octane and decreases engine knock, it has proven to be toxic to humans and eventually will be phased out entirely. Since the 1980s, ethanol has replaced lead in many areas. Also an octane booster, ethanol, most regulators believe, does not pose major health hazards.

According to the Environmental Protection Agency (EPA), although ethanol significantly reduces carbon monoxide emissions, it is less effective than methanol in reducing volatile organic compounds. In fact, it might even increase VOCs.

"Oxy" fuels

Colorado's Air Quality Control Commission began an air pollution abatement program in January 1988 that required the use of either ethanol or methyl tertiary butyl ether (MTBE). As they contain more oxygen atoms than standard gasolines, they are called oxygenated or "oxy" fuels.

MTBE, developed by the Arco oil company, is a petroleum-based product made in part from methanol. When Congress approved the Clean Air Act in 1990, which required states to crack down on auto emissions, it also approved the use of MTBE. As it boosts the oxygen content of gasoline, it causes it to burn more completely. Other so-called oxygenates, such as ethanol, could not be produced in such abundance, according to oil industry officials.

In California, to comply with state and federal clean air requirements, oil companies retooled their refineries at an estimated cost of $5 billion to manufacture reformulated gasoline with MTBE.

MTBE, however, came under attack by many lawmakers because of questions about its toxicity. The oil industry claims that a ban of the additive would result in serious gasoline shortages and large escalations in prices. While clean air experts in California believe that reformulated gasoline containing MTBE has reduced smog they are concerned about reports that it has gotten into ground water in some areas through leaking underground storage tanks. As it threatens to foul drinking water, both Republican and Democratic legislators in California requested detailed health studies and some are asking for an outright ban. It is considered by some authorities to be a carcinogen as it has been shown to cause various types of cancer in mice and rats. It is also suspected to cause birth defects, asthma and other respiratory illnesses. The new reformulated gas also adds to airborne formaldehyde, another carcinogen.

In late 1997, Tosco Corporation, a major oil company, called for the phasing out of MTBE, because they found widespread evidence that it had contaminated lakes and underground supplies of drinking water. The oil company argued that it could make clean-burning gasoline without the use of so much oxygenate but because of regulations it is not allowed to do so.

Motorists and automotive repair shops have reported an increase in auto repairs and even engine fires as a result of MTBE. In the early 1990s, MTBE became an industry unto itself. Suddenly, 20 billion pounds a year of it were produced, making it the second most common commercial chemical produced in the United States.

Benzene is a naturally occurring component of petroleum and has been used in gasoline to boost combustion. It, however, is an even more potent poison than MTBE. In fact, it is so toxic that in 1986 it became the first pollutant ever listed by the Air Resources Board as a toxic air contaminant. Like MTBE, it has been discovered in a number of shallow aquifers near leaking gasoline tanks.

Ethanol is considered less toxic than methyl tertiary butyl ether (MTBE) as a gas additive. It is made from grains, usually corn. According to the U.S. Environmental Protection Agency it can reduce carbon monoxide emissions by 25 percent or more. EPA studies have suggested that the blend may slightly increase nitrogen oxide (NOx) emissions, but it has not been determined to what extent.

Used in the pure form, ethanol reduces most forms of air pollution that come from automobile exhausts. It is a fuel extender as well as an effective octane enhancer in gasoline. Unlike MTBE, there have been no reports of clogged fuel injectors or any other fuel-related maintenance problems from the use of ethanol blended gasoline.

Both in Milwaukee, Wisconsin and in Alaska there was a rash of health complaints after MTBE was introduced. The problems reported were headaches, dizziness, nausea, coughs and skin rashes. The EPA rejected the request by the governor of Wisconsin to suspend the reformulated gasoline. In Alaska, however, when MTBE was replaced by ethanol as the additive, the health complaints disappeared.

MTBE is preferred over ethanol because it is mixed into the gasoline at the refinery and sent out through the pipeline system, whereas ethanol has to be added just before distribution. Both the oil industry and the EPA have pointed out that while exposure to MTBE could result in some health problems in some people, it does reduce ozone and carbon monoxide levels which they believe are much more hazardous to the health.

Weather and Chemistry

Weather is the key to nearly all air pollution problems. In every serious episode of smog, weather has played a part. In London, in Donora and in Los Angeles, smog becomes trapped in a localized area because of a temperature inversion and the lack of air movement. Although regulators can stop factories from spewing out toxic fumes, there is so far no practical way to control temperature inversions or make the wind blow.

The amount of pollutants in the air will determine the harmful effects. Industrial plants that put out poisonous fumes should never be located near populated areas. Toxic substances that escape into the air may not create significant problems for those at a great distance from the source, any more than a drop of ink in the ocean can be detected several miles away.

Ozone is in the stratosphere and there is considerable debate as to whether it can descend into the lower troposphere, eventually reaching the earth in concentrations that could cause health problems.

The ozone experts

The results of a study of ozone pollution were made public in a 1992 report approved by the Governing Board of the National Research Council. Involved in the study were the NRC, the Committee on Tropospheric Ozone Formation and Measurement; Board on Environmental Studies and Toxicology; Board on Atmospheric Sciences and Climate; and the Commission on Geosciences, Environment and Resources. One of the conclusions of this 500-page report titled *Rethinking the Ozone Problem in Urban and Regional Air Pollution* was that ozone control programs have not worked. The report states:

> "Despite the major regulatory and pollution-control programs of the past 20 years, efforts to attain the National Ambient Air Quality Standard for ozone largely have failed....

> "The principal measure currently used to assess ozone trends ... is highly sensitive to meteorological fluctuations and is not a reliable measure of progress in reducing ozone over several years for a given area .... The State Implementation Plan (SIP) process, outlined in the Clean Air Act for developing and implementing ozone reduction strategies ... is seriously flawed in practice because of the lack of adequate verification programs."

Chilean nitrate beds

Over millions of years, weather has built up the Chilean nitrate beds as well as nitrate deposits in other areas. Since nitrates, one of the byproducts of natural smog, are so soluble in water, commercial ventures are restricted to arid areas like Northern Chile. According to a 1981 report by the Department of the Interior, "Geology and Origin of the Chilean Nitrate Deposits" Geological Survey Professional Paper 1188):

> "The wide distribution of the deposits without systematic relationship to topography or rock type can be explained only by atmospheric transport

> and deposition .... The Chilean nitrate deposits probably formed by slow accumulation of the saline components by deposition from the atmosphere of materials from diverse sources."

Before nitrates could be made synthetically, more than a hundred companies were digging it out of the Chilean nitrate beds. Nitrates have been used throughout the world for explosives and fertilizer. Some heavily loaded ships carrying nitrates have exploded in the past causing widespread damage even miles away.

Geographical locations and natural smog

As the greater Los Angeles area in California has received attention all over the world as one that has a severe smog problem, it is mentioned numerous times in this book. It is also an example of an area that has a marine climate which is characterized by mild temperatures with winter rainfall and dry summers. It is cooler in summer than many other cities in the United States. Because of its geographical location, it has more natural smog. However, it does not have the heavy industry that is found in the Eastern United States and Midwest so there is very little industrial pollution.

Santiago, Chile is the same distance south of the equator that Los Angeles is north. It, too, has natural smog. Both Los Angeles and Santiago have the Pacific Ocean to the west and mountains behind them and very little industrial smog.

In the 1950s a direct correlation was found between hazy clouds in the lower level of the atmosphere and smog. The color of these hazes in dry air suggested the presence of nitrogen dioxide (NO_2).

In March of 1954, Francis S. Stewart of F.S. Stewart Associates, in cooperation with Air Research Associates, undertook an extensive chemical sampling program. The presence of nitrogen oxides in the haze clouds was immediately established. A continuing test program revealed that the concentration of nitrogen dioxide ranged from less than one-tenth of a part per million to several parts per million (ppm) of air.

A sampling program showed that there was little connection between the general level of nitrogen dioxide in the air and automobile engine exhaust. Measurable amounts of nitrogen dioxide often could not be found a short distance from busy freeways, even with stagnant conditions and a temperature inversion. During one series of tests, a nitrogen dioxide cloud hung over the entire Los Angeles area and in the hills above Pasadena. At the same time, tests along the freeways disclosed no measurable NO_2 from Pasadena to downtown Los Angeles. The amount of NO_2 produced by automobile traffic was too small and diffuse to measure. This holds true today, even with the great increase in traffic. However, as almost anything that burns gives off nitrogen dioxide, man-made NO_2 could be a problem for those near the source.

Stewart, a noted chemist, publicly announced the widespread presence of naturally formed oxides of nitrogen, including nitrogen dioxide, in the earth's atmosphere. Along the Pacific Coast of North America they were in sufficient amounts to be a major factor in smog. These findings were corroborated by research sponsored by the State of California.

In a paper prepared by Frank M. Stead, Chief, Division of Environmental Sanitation, Department of Public Health, presented at the University of California, Berkeley, in September, 1956, he describes the finding of oxidants (O_3) and oxides of nitrogen (NOx) in remote areas at the earth's surface, both on mountain peaks and at off-shore Pacific Islands. He stated that due to location of the stations, prevailing winds and other factors, these gases could not have come from man-made sources.

Smog scientists who accepted the hydrocarbon theory nearly 50 years ago did not realize that natural ozone could be many times higher than first calculated. Thus they can tell us how to decrease man-made smog, but not natural smog.

Because of the dry air in the mountains and plains in summer, the hazes basically are nitrogen or sulfur oxides or similar substances which do not restrict visibility as much as they do in coastal areas. On the west coasts of both North and South America, Africa, and around the Mediterranean Sea, a combination of moisture and nitrogen oxides or sulfur oxides often forms a thick haze.

Along the coast, fog is often predominant over haze. When the fog clears and the air starts to dry out, a nitrogen dioxide haze is present which is composed of small acid droplets and possibly nitrogen pentoxide which is white. Ozone has a bearing on the kind of nitrogen oxides in haze; it can help in the conversion of the lower oxides to the higher oxides, thus destroying itself.

Although natural ozone smog may have other pollutants mixed in with it, it is generally not related to the type of smog found in London, England, the Meuse Valley in Belgium, and Donora, Pennsylvania.

Weather determines the path and concentration of all contaminants. The greatest influences on the average concentration of natural smog are latitude and topography which also give certain coastal areas a Mediterranean-type climate.

The Los Angeles basin is ringed with mountains except at the seaward side. The San Gabriel mountain range in Southern California acts as a barrier to the cold blasts of air coming down over the interior of the Western United States in winter and helps to provide a mild winter climate.

Santiago, Chile, which has the Andes behind it and the Pacific Ocean to the West also has natural smog.

The marine layer of air which invades the coastal plains throughout the summer keeps the coastal areas of the west side of most continents relatively cool in summer and prevents high, desert-like temperatures from scorching the land. Without the sea air, temperatures in the Los Angeles area in summer, for example, would frequently be over 100 degrees Fahrenheit, as they are in the Imperial and Coachella Valleys of California. This also holds true for other coastal cities where the ocean is to the west.

In a temperature inversion the air temperature is warmer above than below. These temperature inversions occur all over the world. When the sun goes down, cooling begins immediately at the surface of the earth with the most marked effect often within the first few feet above the ground. On clear, still, cloudless nights, increases in temperature on the order of several degrees for the first fifteen feet are not uncommon. As much as a 25-degree temperature difference between ground level and the top of a tall building has been observed.

Unless a pollutant is discharged above the top of an inversion, or has been charged with sufficient heat to pass through the inversion, it will be held beneath it. This condition holds all over the world. It is not unique to Los Angeles.

When air sinks very rapidly it heats up at the rate of about 5.5 degrees F. per thousand feet. This accounts for temperatures of 80 to 90 degrees or more over the Pacific Ocean where there are no land masses to heat. In summer the difference in temperature in the main sea air inversion may range from 10 to more than 30 degrees. The air at the base of an inversion is more dense because it is a lower temperature than the air on top. This prevents much interchange between the lower and higher strata, especially at night.

During the daylight hours, as the sun heats up the earth, it causes the air to rise in columns and, conversely, the upper air to sink further. There is also an increase in the up and down motion of air over rough ground and over rough seas. These factors cause more of an interchange of air between the surface strata and the inversion layer. Anyone flying during the day can often feel the turbulence above the inversion caused by the superheating of the ground. Local contaminants are often held down at night. During the day, however, heating may cause a dispersion through a thicker layer and also bring ozone to the ground.

Fog often occurs along the west side of many continents throughout a large part of the year and sometimes moves many miles inland. All of these are manifestations of a marine climate. It is the winds from the westerly directions and the subsiding air that bring in the great percentage of the naturally formed ozone and nitrogen oxides throughout the year in the coastal and western regions of many continents. High velocity winds off the ocean may keep the interior free of smog for days at a time even though ozone and nitrogen dioxide may be recorded.

A weak and dry low pressure area off Los Angeles, which local meteorologists often refer to as the Catalina Eddy, may be a factor in concentrating natural smog. During the late spring and early summer, prevailing winds sweeping down from the Gulf of Alaska tend to parallel the Southern California coastline. When they reach Santa Barbara, the winds swirl counterclockwise following the curve of the coast. They have been named the Catalina Eddy because they are often near or over Catalina Island. This dense overcast, so prevalent in the area at the beginning of summer, is what those who live along the coastline call the "June gloom."

One of the errors in smog research is the notion that air pollution coming in across a coastline throughout the day during a smog siege is the same air pollution that was blown out to sea the night before. Although offshore breezes will carry a certain amount of air pollution out to sea, for much of the year the velocities are very light. Ozone and some other contaminants are quickly destroyed at the surface at night and by moist air.

Even before the industrial age, before the use of coal, before Homo sapiens began polluting the atmosphere, there were natural pollutants. As was stated at the beginning of this book, Native Americans and early explorers noticed the hazy gray or brown air that appeared along the coast and settled in coastal valleys in Southern California years before there were automobiles and factories.

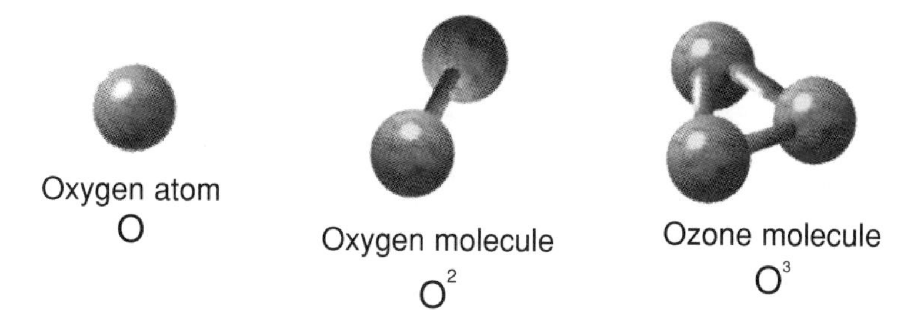

When a third oxygen atom bonds to an oxygen molecule, ozone is created.

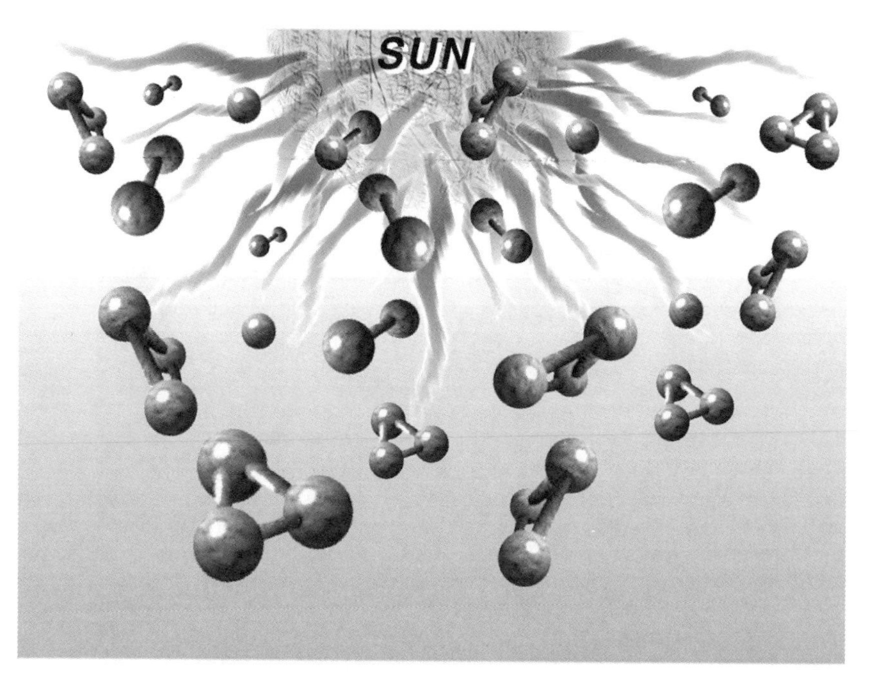

In the stratosphere, energy from the sun is strong enough to split some of the oxygen molecules, forming ozone.

Downtown Pittsburgh, Pennsylvania before and after anti-smoke laws went into effect.

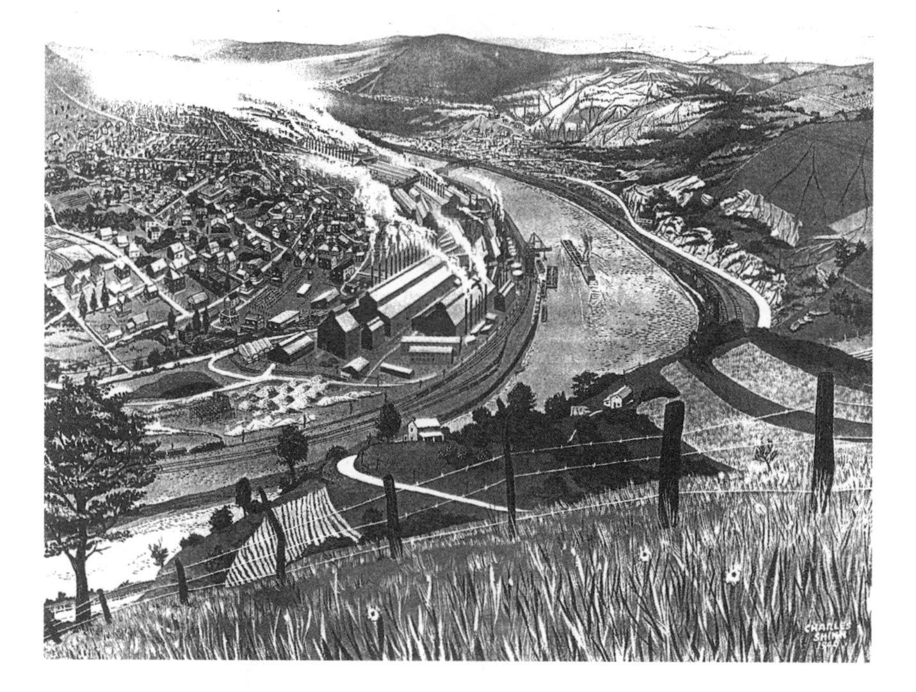

In Donora and Webster, Pennsylvania, the Monogahela River valley is narrow and lies between steep, sloping cliffs. In late October 1948, toxic smoke and fumes from factories and railroad engines combined with fog and became trapped in the cold, motionless air. Half of the 14,000 inhabitants became ill; 20 people and nearly 1,000 animals died.

Volcanic eruptions spew gases, ash, dust, molten lava and solid fragments into the air. Extremely large quantities of sulfur are emitted by some volcanoes. They form natural aerosols that contribute to the depletion of the ozone layer.

Lightning produces large amounts of ozone and nitrogen dioxide.

Home incinerator used throughout the greater Los Angeles area prior to 1958 to burn household trash.

Orchard heater, also known as a **smudge pot**, used to prevent frost damage in citrus and avocado groves.

The Health Effects

With the enormous sums of money spent in combating ozone smog, it could be concluded that ozone is a lethal air pollutant threatening a huge segment of the population. However, there are differences of opinion within the scientific community as to the extent of the hazard. Although most researchers agree that exposure to high concentrations of ozone over a prolonged period of time can cause burning eyes and difficulty in breathing, studies have found that most healthy people recover completely from the effects within a matter of hours.

For people with asthma, emphysema and other respiratory diseases, however, ozone exposure can be far more damaging and the harmful effects can last much longer.

The effects of ozone on human health have been studied by scientists for years. In controlled experiments in test chambers with pure ozone, they have found that in certain concentrations ozone can cause chest pains when breathing deeply, shortness of breath and coughing. For most people, there are three factors which affect their reaction to ozone. They are (1) the amount of ozone in the air, (2) the length of time they are exposed to it and (3) how deeply they are breathing.

In an extensive Congressional Office of Technology Assessment (OTA) report in 1989, the medical literature on the health effects of ozone was reviewed. The conclusions were that even relatively high ozone levels do not cause long-term health problems for healthy individuals. Different studies have produced different results and conclusions, however.

More recent studies have been made of ozone levels in the real world. One study by researchers at the Nelson Institute of Environmental Medicine in New York and the Environmental Protection Agency examined the records of hospital admissions during very smoggy days in hospitals in Toronto and New Jersey. In New Jersey, they reported, there were on average an additional 1.07 emergency visits during days when the ozone level was higher than .06 ppm. The studies could not, however, specifically isolate ozone and some of the effects may have been caused by other pollutants in the air.

In a study by the National Institute of Environmental Health Sciences, a United States government agency, to determine whether ozone might cause lung cancer, rats were exposed to ozone levels more than eight times as high as the federal standard. After two years of such exposure, they found no evidence of cancer in the rats.

According to a 1993 report by the Nelson Institute, Morton Lippman of the Institute of Environmental Medicine at New York University found that, "Successive days of exposure of adult humans in chambers of ozone lead to an adaptation of lung function." People got used to the ozone and built up a tolerance for it.

In a study by the independent Health Effects Institute of Cambridge, Massachusetts, researchers found that laboratory rats exposed to high levels of ozone for 20 months had "little or no measurable impact on lung function." The study indicated that the rats' lungs apparently adapted to the ozone, which protected them from damage. The authors of the study concluded, "Evidence from previous animal and human studies supports this suggestion that prolonged ozone exposure may lead to some degree of tolerance."

The January-February 1993 issue of the Archives of Environmental Health reported on a study of the "Effect of ambient ozone on peak expiratory flow [exhaled breath] of exercising children in the Netherlands." The authors stated, "The most important finding of this study was the lack of association between children's peak flow difference before and after training sessions and ambient ozone concentration during that training."

There is clearly more than one opinion in the ozone issue. In a 1990 report, "National Air Quality and Emission Trends" produced by the U.S. Environmental Protection Agency, it states, "The health effects of ozone are varied and severe. Because it is a reactive oxidizing agent, ozone tends to attack cells and break down biological tissues. It can be particularly damaging to lung tissue, even at low concentrations."

The American Lung Association believes ozone can trigger asthma attacks. The South Coast Air Quality Management District in the Los Angeles area states that ozone exposure also can reduce resistance to infections.

The mixture of ozone and nitrogen dioxide is toxic

The most harmful constituent of natural smog is the combination of nitrogen dioxide and ozone. These concentrations often exceed the maximum allowable standards set by the U.S. Public Health Service. In the field of industrial hygiene a mixture of ozone and nitrogen dioxide is extremely toxic in very low concentrations.

Particulate matter

Many health regulators consider particulate matter California's deadliest air pollutant because the tiny particles can escape the body's defense mechanisms and penetrate deeply into the lungs. PM 10 and PM 2.5 have been linked to chronic respiratory disease and premature deaths from heart attacks, pneumonia, asthma and other ailments.

According to a study by the Harvard School of Public Health, 15% to 17% of people who live in polluted cities are more likely to die prematurely than those in cities with the cleanest air. The fine airborne particles of smoke, soot and sulfate which were seen in the study were not the type of pollution that was regulated under the clean-air standards. A number of studies have been undertaken to determine the health effects of particulate matter. Two types of epidemiological studies were considered by the EPA. One was a short-term acute mortality study comparing daily particulate matter levels and mortality in several different locations in the U.S. The other study was of groups of individuals over a number of years in various locations. Mortality rates in a given location were related to the annual average PM 10 or PM 2.5 concentrations. In this study the rates were adjusted for smoking and other variables. In both of these studies there was shown to be a positive relationship between annual mortality and particulate matter. They determined that the long-term effects were more damaging than short-term exposures.

There was a third study which was considered by the EPA which showed no relationship between mortality and particulate matter. The EPA dismissed this study partly because it had a smaller sample size.

PM 2.5 was found to be much more damaging than the larger particles within PM 10s. According to Morton Lippmann, "PM 2.5 is a better measure than any alternative metric of the complex in the particulate mass that is causing excess mortality and morbidity." There is a stronger and more consistent association between respiratory diseases and ultrafine particles than with the larger fraction of PM 10 particles.

Very tiny particles are often contaminated with other pollutants such as ozone, sulfur dioxide and carbon monoxide. Therefore, the culprits may be those chemicals as much as or instead of the particles. Hot weather and humidity must also be taken into account when evaluating hospital admissions during smoggy seasons. There will undoubtedly be many more studies before a consensus is reached, if it ever is, about just what our smog consists of and how harmful it is.

During the past three years several articles have appeared in British journal *The Lancet* regarding the effects of nitrogen dioxide (NO_2) on asthmatics. The highest levels of NO_2 are found in the home, especially in kitchens with gas stoves where it can be ten times the amount found in the outside air. One study reported in *The Lancet* found that in very high concentrations it could aggravate respiratory symptoms, especially for those who spent an hour or more in a kitchen during the cooking of a meal on a gas range.

Another study found sulfur dioxide to have a greater effect than NO_2 on health. But the combination of the two gases were considered more harmful to health than the individual constituents.

The California Air Resources Board states in their pamphlet, "Facts About Air Pollution and Health," that nitrogen dioxide is "an irritating gas that may increase the susceptibility to infection and may constrict the airways of asthmatics."

There are many things in the air we breathe that can be damaging to our health. Both our indoor and our outdoor environment need to be monitored. For some people, dust mites in the home can cause allergic reactions as well as glues and fibers in wall-to-wall carpeting and animal dander. Drifting pollen, found mainly in spring and autumn months, creates havoc for hayfever sufferers.

Probably the most hazardous constituents of the air one could encounter outdoors would be toxic fumes from factories. The industrial accident in 1984 at the Union Carbide plant in Bhopal, India which killed more than 3,300 and caused over 20,000 illnesses was a tragic example of that.

In 1952, over 4,000 Londoners died during a four-day smog attack created by coal smoke and sulfur oxides from huge power stations, factories and industrial plants as well as from home furnaces.

Indoors, the main cause of air pollution is tobacco smoke. Household smoking increases the risk of hospitalization for respiratory illness by 55%. According to a 1996 report from the Centers for Disease Control, second-hand smoke causes about 350,000 cases of bronchitis, 430,000 cases of asthma and 152,000 cases of pneumonia in children annually. Between 136 and 212 children under the age of five die each year due to second-hand smoke.

Second-hand smoke kills over 53,000 people per year. The Environmental Protection Agency declared in 1993 that secondhand smoke causes about 3,000 cases of lung cancer annually. Before smoking was banned on airlines' domestic flights in 1989, it is estimated that 60,000 flight attendants developed ailments such as cancer, heart disease, bronchitis and asthma from the cigarette smoke of airline passengers.

In late 1996, the Journal of the American Medical Association reported on research that found a link between smoking and breast cancer. The rise in the breast cancer rate in the past few decades appears to be due to increased smoking among women, according to researchers from the National Cancer Institute and the National Center for Toxicological Research.

Finally recognized as a serious health hazard, smoking is banned in public places throughout much of the United States. In California, smoking is no longer allowed in restaurants, bars and almost all workplaces.

There are several thousand constituents identified in tobacco and tobacco smoke. Some of the chemicals in tobacco are acetone, methanol, nicotine, cyanide, ammonia, formaldehyde, carbon monoxide and acetylene.

Smoking and exposure to second-hand smoke are major health hazards, killing literally millions of people throughout the world every year. All the deaths from all other forms of air pollution in the world make up only a small percentage of those who die each year as a result of smoking.

The History of Smog Regulations

In thirteenth century London, bituminous coal became the primary source of fuel when the forests became depleted from over-harvesting. As the scarcity caused the price of wood to soar, coal became popular with local craftsmen and manufacturers. In the winter, the black smoke from coal fires mixed with the fog. Temperature inversions held the smog close to the ground and the inky air permeated the city and everything in it.

While the coal trade was still in its infancy, the local residents found it offensive. They complained of the smoke's disagreeable odor and considered it very detrimental to health. By the late thirteenth century, in what was to be the British government's first attempt to control air pollution, a proclamation was issued in the name of Edward I, drastically curtailing the use of coal in the city. In 1306, in what is believed to be the first official penalty for polluting, a factory owner caught disobeying the Royal Proclamation was tried, found guilty and beheaded. However, the bituminous coal which was abundant and cheap, continued to be used by London's artisans and manufacturers in defiance of the order.

In the 1400s the poorer classes began burning coal instead of wood in their homes for cooking and heating. By the 1600s, wood had become too expensive even for the wealthiest London families and, in spite of its offensive reputation, coal came into wide usage in all homes for cooking and heating purposes.

In the 1800s Charles Dickens referred to the murky London air in both *Tale of Two Cities* and *Oliver Twist.* But it wasn't until the early 1900s that the House of Commons introduced their Smoke Abatement Bill. It was withdrawn and in spite of the fact that they tried several more times to address the problem of London's pollution, they were unsuccessful.

No one wanted to place curbs on the industrial polluters. It was simply not economically feasible. As a 1921

Final Report by the Minister of Health's Newton Committee stated:

> "The introduction of legislation which might prejudicially affect important industries is quite out of the question....
>
> "The burning of raw coal is a dirty, wasteful and unscientific practice, and on the grounds of economy, as well as of public health, it should be restricted as much as possible but... after full consideration, we do not consider it practicable at present to propose legislation dealing with smoke from private dwelling houses."

In 1926 the British Parliament passed the Smoke Abatement Act. Nothing was done, though, to limit the smoke and soot from domestic chimneys or the black smoke from industrial stacks, so the air pollution continued to worsen.

The California Smog Problem

In the 1890s the discovery of oil stimulated expansion in Southern California. The development of the motion picture industry in the early 1900s lured still more people into the area. Because of its mild climate, it became a flourishing tourist destination and the home of many retirees.

The depression era of the 1930s, compounded by years of drought and dust storms in the Midwest, brought desperate migrants from New Mexico, Oklahoma, Arkansas and Texas, hoping to find enough work to feed their hungry families. But it was during World War II that the numbers of people living and working in the Los Angeles Basin changed most dramatically. From all over the United States people poured into California to find jobs in defense plants which had blossomed throughout the area.

The population exploded with the young blue-collar workers eager to contribute to the "war effort." At the same time, they were drawn to the palm trees and balmy ocean breezes of Southern California with dreams of making their fortunes and raising their families in new, enchanting surroundings.

The new citizens were mostly people from the heartlands of the country, with a strong American work ethic and a deep sense of patriotism. And they were productive. In the first 200 days of 1945 alone, over 240 Victory cargo ships were launched from the Calship Yards in Los

Angeles.

The War brought many large, new industrial plants and plentiful jobs. Soon there was an urgent need for housing and schools; streets and highways. Along with the congestion came an increase in building. Food, clothing and furniture stores popped up everywhere. New factories discharged all kinds of evil-looking fumes. Trash and industrial waste were burned in open fires or primitive incinerators. The smoke from smudge pots in citrus groves and the burning of the remains of past harvests in agricultural areas filled the air.

As traffic increased there were the exhaust fumes of automobiles, trucks and buses. Lumber yards, fertilizer plants and coffee roasting operations all contributed to the "bad air."

It seemed the increased land development and industrialization during the war years had caused an increase in air pollution. Certainly there was more smoke; there were more particulates in the air than before. During some periods of hot weather in late spring, summer and early fall, however, the air quality seemed worse. According to newspaper accounts in the early 1940s, Los Angeles began experiencing days when smoke and fumes blanketed the city as well as the suburbs and large numbers of people retreated into air-conditioned buildings to escape the smoggy air.

On September 8, 1943, a toxic brown carpet of smog covered not only the Los Angeles basin but extended up into the surrounding foothills. Citizens, experiencing stinging, watery eyes and pains in their chest became enraged and demanded that public officials do something to clean up the air. At that time most of the acrid air was thought to be solely the result of industrial pollutants.

A synthetic rubber plant which was owned by the Southern California Gas Company was blamed as the primary polluter and was shut down until emission control equipment could be installed. There was very little improvement in air quality, however, even while the plant was shut down or after the emissions had been controlled.

At that point it should have become clear that major cause of the irritating smog was not the result of any particular industry. The noxious, hazy air often invaded regions far removed from any industrial or automotive emissions. Public officials who were trying to placate the public continued to search for a culprit.

The objective of the regulators was to return to the

level of air quality in Los Angeles that had existed prior to the 1940s, when the air was perceived to be clean. However, there were no measurements of air quality back then. They could only make estimates and act on their assumptions.

The public's concern over the air quality in the Los Angeles Basin became a timely subject for local newspapers. There were a number of editorials over the years in the Los Angeles Times newspaper about smog. In 1947 The Los Angeles Times brought in Professor Raymond R. Tucker, the former smoke regulation commissioner of St. Louis, Missouri, to study the Los Angeles air. Professor Tucker had been successful in greatly reducing the severe smoke pollution in St. Louis. The L. A. Times published his report on the front page. In it, the professor noted the rapid industrialization and population expansion in the Los Angeles area. He investigated several major sources of air pollution and reported on the noxious fumes which were being released into the air. Tucker identified these as heavy industries, foundries, backyard incinerators and smudge pots and recommended the imposition of stringent controls upon them.

He did not, however, indict automobiles as a substantial cause of the air pollution, noting that there had been far fewer vehicles entering the area during the war years and it was during that time that air quality had been the worst. Although there were numerous theories regarding the cause of the noxious smog problem, he stated, no one theory had been proven. He recommended further research.

It was also in 1947 that the Western Oil and Gas Association, fearing undue regulation of its member companies, contracted with the Stanford Research Institute (not a part of Stanford University) to study the smog in the Los Angeles area. Over the next few years, although they completed three separate reports identifying specific chemicals in the air as causing serious eye irritation, they did not identify any particular chemical as the "mysterious pollutant." It was to be several years before the pollutant was identified as ozone.

In an article in The New York Times on November 11, 1949, Dr. Louis McCabe reportedly placed the blame for the persistence of noxious "smogs" in many American cities on official mismanagement, misguided public enthusiasm and selfish industrialists. McCabe was chief of the Bureau of Mines office of air and stream pollution prevention research at the time. He had helped organize the Air Pollution Control Authority in Los Angeles and became the first director of the Air Pollution Control District. He believed that citizens had been deluded by an extensive "folklore" about smog and industrialists had thwarted campaigns against it.

Dr. McCabe was quoted as saying that the effort against air pollution was still a failure "because industry believed that air pollution control costs too much." He claimed:

> "There were 'cooperative' programs with the dual objectives of delay and defeat. Engineers were assigned to write diverting papers on the minutiae of the problem, and trade journals editorialized on the unreasonableness of 'do-gooders.' These tactics haunt the sincere efforts of progressive industry today."

McCabe's experience in regulating "smoke-stack" industries in the eastern United States did not apply to the type of pollution seen in Los Angeles County. He soon became the target of criticism from citizens who felt that he wasn't doing enough to clean up the air, and by industry for doing too much. Harassed from both directions, McCabe resigned in 1949 and Gordon Larson replaced him as Director of the Los Angeles County Air Pollution Control District.

With the fear growing among the citizens and the press that the Los Angeles haze would turn into a London-type "killer" fog, Governor Goodwin J. Knight took action. In 1953, he appointed an air-pollution review committee. The committee, chaired by Arnold O. Beckman of Beckman Instruments, proposed several possible solutions to reduce the pollutants over the short term. They were (1) reduce hydrocarbon emissions by improving the procedures for transferring petroleum products; (2) set standards for auto exhausts; (3) encourage trucks and buses to burn liquefied petroleum gas instead of diesel fuels; (4) consider slowing the growth of industries that pollute the area heavily; and (5) ban the open burning of trash.

Backyard trash incinerators were banned in 1958. Many people were opposed to the ban, realizing that burning their trash was a great deal less expensive than paying to have it hauled away. Besides, over 300,000 families owned incinerators.

Other solutions were offered at the time, such as developing better incinerators or installing emission control devices on existing private incinerators similar to what had been done on factories. It was suggested that burning of trash could be limited to specific days of the week in each neighborhood to limit the amount of smoke in the air at any one time and done only in the afternoon. The U.S. Weather Bureau could ask citizens not to burn when a low inversion was predicted.

Over the long term, the Beckman committee hoped that Los Angeles would develop a program to control auto emissions, develop a rapid transit system and begin a cooperative program to regulate industrial pollutants. At the very time that the report was issued, however, the area's public train system was being dismantled. The citizens of Los Angeles, who had enjoyed the low-cost and convenience of an excellent public transit system for years, now had to rely on private automobiles for transportation. Many felt that it was a great disservice to the public. However, it turned out to be a boon to auto manufacturers and oil companies who, as a result, sold more of their products.

Some public officials insisted that Californians were a

different breed of citizen. They were ruggedly independent people who had no use for public transportation. They wanted their own cars so that they could come and go as they pleased. Even though there was an outcry from many citizens who could not afford to own their own cars, their voices went unheard. In Los Angeles County, the "red cars" were scrapped and the rails were dug up or paved over.

The industrial polluters put up much more resistance than the public. It has been difficult, historically, to control toxic industrial emissions because of the need to balance environmental interests with a region's economic needs. Most industries, however, did submit to requirements that they use vapor-recovery equipment when they transferred petroleum products. Most even put floating roofs on their storage tanks so there would be no room for vapors to accumulate over the toxic liquid. It would be several more years, though, until vapor recovery equipment would be required on gas pumps.

The Beckman committee's recommendations eventually grew into an air-quality management plan but it was a slow process which was reshaped many times. Los Angeles County's neighbors, Orange, Riverside and San Bernardino Counties soon began their own pollution-control programs. All four counties, in subsequent decades, have experienced an explosive growth in population and vehicular traffic. It became apparent that, as smog doesn't respect political boundaries, they would have to coordinate their efforts.

During the 1960s regulations were adopted to eliminate industrial solvents. A wide variety of businesses, such as manufacturing, construction and dry cleaning were affected by the rules. Most of them changed their solvents rather than having to install expensive new equipment. It was a hit or miss proposition, however. Many of the new solvents either did not adequately do the work they were intended for or, in some cases, they created new environmental problems and, as usual, the costs increased.

The regional governments of Los Angeles, Orange, San Bernardino and Riverside Counties tried in 1975 to voluntarily consolidate their pollution-control programs but were unsuccessful. Therefore, two years later, the California legislature forced an alliance among the local programs by creating the South Coast Air Quality Management District, referred to as the SCAQMD. The new

District was given jurisdiction over the four Counties, an area of 13,350 square miles.

The SCAQMD was initially responsible for stationary sources of air pollution. The California Air Resources Board in Sacramento regulated mobile sources such as cars, trucks and buses. Adopting an attitude held by most business leaders, that industries were already about as clean as they could get, the regulators targeted automobiles as the major polluters and thus it was the public who paid.

In 1979 and 1982 the SCAQMD adopted air-quality management plans. The federal Clean Air Act, at that time, was requiring all American cities to achieve federal standards by 1987. But it was generally agreed that the task would be impossible in Los Angeles.

As the deadline for compliance drew near, environmentalists and even some business leaders attacked the SCAQMD as being too lax in their enforcement. Therefore, in 1987, the governing board of the SCAQMD was restructured and the legislature granted it broad powers. The SCAQMD is now responsible for accomplishing local, state and federal clean air goals.

According to an article by James M. Lents and William J. Kelly of the SCAQMD in *Scientific American* (October 1993), man-made air pollution in California has largely been cleaned up. The authors of the article report that Southern California has made great progress in the fight against air pollution. Motor vehicles and industries in California are now among the cleanest in the world. New cars today create only about a tenth of the pollution as did the cars sold in 1970, they reported.

Steven D. Mazor, Principal Automotive Engineer for the Automobile Club of Southern California, says that newer cars are very clean. The only exceptions are those cars whose catalytic converters have been tampered with or are defective. The regulations have been much stricter on automobiles than on trucks or industry. Whereas in 1965 autos were said to have created half of the pollution, now they contribute less than 25% to air pollution.

The air pollution problem in New York City

New York City has actually had a longer history of air pollution control than Los Angeles. During the early 1900s, the New York Sanitary Code prohibited excessive emissions of smoke, cinders, dust and fumes. In the 1930s the Department of Health's smoke control unit was charged with trying to enforce air pollution regulations. After World War II, the Bureau of Smoke Control was created within the Department of Housing and Buildings. It was ineffectual, however, and city officials realized that controlling smoke was not enough.

In 1952 a separate department was created. The new Department of Air Pollution Control was given the power to license equipment that created objectionable emissions, and to require them to alter the equipment to comply with regulations.

In 1965 the City Council, responding to pressure from the U.S. Department of Health, Education and Welfare and Citizens for Clean Air, a special Committee on Air Pollution was formed. A short time later the Committee issued a report noting that suspended particulates in the New York air were higher than any other major U.S. city. They found that the annual city-wide dustfall averages had changed very little since 1952. They averaged 60 tons per square miles per month. In Manhattan, they were 80 tons per square mile per month. Dust was not the only problem, however. They also discovered that the sulfur dioxide values were also higher in New York than any other major city in the country.

The City Council's Committee on Air Pollution reported that the most significant sources of air pollution in the city were not heavy industry, as there was not much heavy industry in New York City. The Committee didn't target automobile emissions either. Industry and automobiles took a back seat to the major source of pollution, which was believed to be the combustion of bituminous coal and residual fuel oil.

In 1966 a bill was introduced calling for the reduction of sulfur dioxide emissions from fuel burning, the elimination of bituminous coal use, the upgrading of incinerators and the banning of refuse incineration in new buildings. It was Consolidated Edison, a large public utility, that was burning large amounts of the fuel consumed in the city and was contributing a large percentage of the sulfur dioxide in the atmosphere. The proposed bill required the use of low-sulfur fuels and the upgrading of incinerators through the installation of emission control equipment. The New York Times labeled the bill the most stringent air pollution legislation in the country.

The High Cost of Air Pollution

Worldwide pollution of our land, our water and our air is a very real threat and we do not have infinite resources to fight it. Toxic fumes from industrial polluters are still making people sick all over the world. In some areas they have caused immediate illness and even death to people living near factories. Many may die months or years from now as a result of toxic fumes they are exposed to daily.

On very smoggy days our eyes may burn, we may have a dry throat and feel short of breath. We may have a headache and nausea as well. This isn't our imagination; something is definitely going on. There are elements in smog that can cause us to feel tired and even harm our health. For example, carbon monoxide fumes that emanate from old cars can impair visual perception and manual dexterity. It can make a person feel drowsy and tired, reducing their learning and driving abilities. It is a well-known fact that in enclosed spaces it can and does kill.

Particulate matter can contain any number of toxic substances, many of which have yet to be identified. Some of them are natural, some are man-made. These fine particles can cause painful breathing, lung damage and even contribute to cancer. Farmers, ranchers and road workers have known about the problems of blowing sand, dirt and dust for centuries as they may have become ill from the fine particles they have been forced to breathe in the course of doing their work.

Most of this particulate matter is of natural origin. The principal protection these workers have at their disposal are masks that filter out some of the dust. Whenever possible, they can wet down the soil so that it doesn't blow around. Beyond that, there isn't a great deal that can be done.

There are many kinds of air pollution created by human activity all over the world. Most of it can be controlled and some of it has already been addressed. Smoke from home incincrators has been eliminated. Lead has been removed from gasoline. Emission devices have been installed on autos so that they now emit far less carbon monoxide than before.

Although they have already cost billions of dollars, most of the relatively inexpensive pollution control measures were implemented during the last 20 years. Future reductions in emissions are likely to be far more expensive than earlier ones. In the early 1990s, major changes were made to the U.S. Clean Air Act. It called for electric power plants to make sharp reductions in emissions of sulfur dioxide from the petroleum products they burned. It required most major sources of hazardous air pollutants to install state-of-the-art emissions control equipment.

The Clean Air Act also enacted a number of new measures in its quest to improve air quality in areas where the national ambient air quality standards (NAAQS) were being violated. This problem, however, is by far the most difficult to solve, according to Dr. Alan J. Krupnick and Dr. Paul R. Portney of Resources for the Future, a Washington D.C. based nonprofit research organization.

The Environmental Protection Agency (EPA) stated in 1989 that more than 66 million people in the United States lived in counties where the ozone standard was being exceeded at one or more monitors.

The Office of Technology Assessment (OTA) released a major study of air quality problems in the United States in 1989. They estimated that by the year 2004, control measures would reduce total annual emissions of ground-level ozone resulting from the control of volatile organic compounds (VOCs) from 11 million to 7 million tons, a 35% reduction. However, areas such as Los Angeles were predicted to remain in violation even after controls were implemented.

The annual cost associated with their ambitious measures, according to the OTA, would be $6.6 billion to $10 billion in the nonattaiment areas alone. (Nonattainment areas are those where no appreciable improvement has been made.) When they added in the costs that would be borne in the attainment areas (areas no longer in violation) the estimated total skyrocketed to between $8.8 billion and $12.8 billion per year.

In their estimates, they did not state the cost of the mandatory introduction of alternative fuels (methanol, ethanol or reformulated gasoline) that was called for in the new Clean Air Act amendments. This adds at least $3 billion annually. There is also no mention of yet another round of vehicle emissions reductions that will probably be required under the amendments which will add an additional $5 billion per year.

In Southern California, the SCAQMD outlined a Smog Plan whose costs will be astronomical. A study prepared by the National Economic Research Associates, Inc., of Cambridge, Massachusetts, contains an estimate of the cost of the Plan for the Los Angeles Basin and surrounding areas as $13 billion dollars per year.

The SCAQMD's 1994 Air Quality Management Plan called for an emission-based Vehicle Registration Fee. According to Thomas McKernan, President and CEO of the Automobile Club of Southern California, in an article in the September/October 1994 issue of *Avenues,* the Automobile Club magazine, it could increase annual registration fees by up to $1,000 per vehicle. An At-the-Pump-Fee could increase fuel taxes up to $2 per gallon and Congestion Pricing could impose a 15-cents-per-mile fee for driving on roadways designated as congested, he claimed. In his article he stated:

> "These measures are intended to reduce travel by imposing financial hardships on the same Southland motorists who have been paying for the

> clean-air gains we've made and are continuing to make. They are examples of public policy based upon the misperception that cars are, and will remain, the most significant contributor to air pollution. They're not only unnecessary but also will fail to produce the gains in air quality we all desire. That's inequitable as well as onerous."

The U. S. Congress is rethinking some of the environmental laws that have been enacted during the past 25 years. Clearly, there has been a lot of wasted money, a lot of errors and not a lot of science involved in many of the regulations. In the mid-1990s, John D. Graham, a professor of public policy at Harvard's School of Public Health, told Congress that we're ignoring some large, documented risks while regulating some nonexistent ones. He stated that it's "paranoid and neglectful at the same time."

Paul Portney estimated that under the 1990 Clean Air Act urban air pollution regulations are costing big and small businesses $20 billion annually but only saving $12 billion or less in health costs.

Since 1992 the SCAQMD has forbidden repair shops in the Los Angeles Basin to add R-12 refrigerant (Freon) to a leaking air-conditioning system. It is against the law to put Freon in the air conditioning systems of older cars that require it because Freon is a type of fluorocarbon which some experts believe damages the Earth's ozone layer. There is an authorized substitute, known as R-134A which reportedly does not damage the ozone layer. However, it can't be used in a system designed for R-12 without making costly modifications.

Motorists will need to find a competent air-conditioning shop which has equipment capable of identifying leaks as small as one-half ounce per year. That equipment costs $1,800 to $4,000 so it probably won't be found at most local gas stations.

Some scientists believe that chlorofluorocarbons (CFCs) and, particularly Freon, do not significantly destroy ozone aloft. One of these scientists is Sallie Baliunas of the Harvard-Smithsonian Center for Astrophysics. As one of the world's leading experts on the physics of the sun, she has recently laid her impeccable credentials on the line by opposing the ban. Early in 1995 she told the Arizona Legislature that its residents would be in no physical danger if the state refused to abide by the ban.

Opponents of the ban on CFC's worried that the proposed alternatives might carry other risks. This concern was borne out in 1997 when scientists reported the first confirmed cases of liver damage among workers who were accidentally exposed to HCFC-123 and HCFC-124. The British medical journal, the *Lancet,* carried the article on the study of these alternatives to chlorofluorocarbons. The study stemmed from an epidemic of liver disease in 1996 at a smelting factory in Belgium. Nine of the workers were found to have acute hepatitis over a period of four months. It turned out that a plastic pipe in the plant's air-conditioning system had leaked.

A 1987 treaty, the Montreal Protocol, is phasing out the use of CFCs in many countries at a cost expected to run into the billions of dollars. Many authorities who oppose the ban see the ozone threat as a myth generated by big government and greedy scientists. Dr. Baliunas worries about the corruption of science which troubles her far more than the perceived threat of ozone depletion. She states that her concern is, "working in a system that's potentially corrupt, that won't look at the facts."

Dr. Baliunas reported in the mid-1990s that there had been little effort to measure the level of ultraviolet radiation reaching the ground. The ozone layer, she stated, is affected by many natural events, including volcanoes and seasonal weather patterns. She explained, "There's a hundredfold difference in just seasonal swings compared to what the man-made depletion is."

In 1997, the EPA proposed strengthening air-pollution standards again. The proposal was met with strenuous protest from industries that contribute the most to air pollution such as oil, utilities and heavy manufacturing. They predicted a loss of thousands of jobs if they were forced to comply with the new regulations. Congressional opponents of the stringent air-quality standards proposed by the Clinton administration pushed for a four-year delay to study the need for the tougher standards. The proposal established standards, for the first time, for some of the smallest particles of soot from burning coal and oil. It also established lower levels at which ozone is considered unhealthy. Among the objections were that the standards were not supported by solid scientific evidence and that implementation would be extremely costly in terms of restrictions on power utilities and industry.

The health costs of air pollution cannot be taken lightly. It is estimated that it kills about 50,000 Americans each year from heart disease, asthma, stroke and various lung ailments. These estimates are based upon studies comparing deaths and hospital admissions with air pollution. Joel Schwartz, associate professor of environmental epidemiology at the Harvard School of Public Health states, "The biggest increase we find on high air-pollution days is in people dead on arrival at the hospital."

The costs vs. the health benefits

In the early 1990s a study was done by the American Association for the Advancement of Science assessing the costs and benefits of trying to control smog. The conclusions were published in an article entitled "Controlling Urban Air Pollution: A Benefit-Cost Assessment" in *Science* (Vol. 252, April 26, 1991) and excerpted in *Consumer's Research* in August 1991.

Ozone

According to the article, the Los Angeles area would remain in violation of the allowable ozone levels by the year 2004 even after the costly controls were implemented. The scientists concluded that the estimated benefits of removing lead from gasoline were well in excess of the costs.

The reverse is true with national ozone control, they reported. The costs of controlling ozone compared with the benefits which would be derived from such controls are unfavorable.

Factory emissions

There are many indications that air pollution is hazardous to our health. In one study that came about serendipitously, scientists were able to conduct a "natural experiment" when a steel mill shut down near Provo, Utah due to a labor dispute. It was the major source of air pollution in the isolated mountain valley.

During the 13 months it was not operating, admissions to the three local hospitals were counted. When they compared them to the admissions for respiratory illnesses before and after the shutdown, the results were striking. There had been 40% more admissions for bronchitis and 17% more for pneumonia when the mill was running.

Particulates

Particulates which are much tinier than ordinary dust can be inhaled into the deepest recesses of the lungs causing illness and death. In a so-called Six City study, researchers from Harvard and Brigham Young universities compared the health of 8,000 adults with local fine-particle counts over a period of 15 years. They found that residents of Steubenville, Ohio, the city with the most particulate pollution had a 25% higher rate of mortality than those of the cleanest city, Portage, Wisconsin.

Almost all of the excess deaths were from lung cancer and heart disease. In a study using larger numbers—500,000 adults in 151 communities, the results were similar.

Smoking

Smoking kills hundreds of thousands of people each year in the United States alone. It has been estimated that smoking accounts for $65 billion per year in health-care costs and loss of productivity.

The health consequences as well as the costs are far greater for smoking than for any other form of human-caused pollution. This fact becomes very clear when one compares the estimated 50,000 Americans who die from air pollution each year with the 434,000 who die from smoking and the 53,000 nonsmokers who die from second-hand smoke.

Billions of dollars have been spent on combating ozone pollution while the money spent on tobacco education and regulations has been in the range of millions of dollars.

Volatile Organic Compounds

The Office of Technology Assessment estimates the cost of a 35 percent reduction in volatile organic compounds (VOCs) would be between $8.8 billion and $12 billion. But the acute health improvements that are predicted as a result of the reduced VOCs have been valued at no more than $1 billion annually and possibly would amount to only about $250 million. The $1 billion estimate is based on the Environmental Protection Agency studies on the health effects of ozone and other photochemical oxidants.

Coal and oil

The recent change in regulations would have a major impact on coal- and oil-burning industries, the utilities and diesel vehicles which were under-regulated before. The EPA estimates that adhering to the new air pollution standard could save about 15,000 lives a year and avoid 8,000 hospitalizations.

The agency claims that the new rules would cost less than $9 billion annually but the health benefits would be worth an estimated $120 billion.

Industry estimates the cost at around $60 billion and with far fewer health benefits.

What We Can Do About It

Both man-made and natural air pollution are made worse by meteorological conditions. Pollutants at ground level are trapped by temperature inversions and motionless air. In the 1950s a number of solutions were proposed to the Air Pollution Foundation in Los Angeles, California. Several of these proposals to clean up the air appeared in an article in the journal *Science* (Vol. 126:637-45, October 4, 1957) titled, "Weather Modification and Smog." In retrospect, many of these "solutions" appear preposterous and even humorous.

While using Los Angeles as an example, the article explained that the sub-tropical inversion is characteristic of similarly situated areas throughout the world.

Most of the early proposals for meteorological modification were concerned with raising or eliminating the inversion, moving the air rapidly across the basin, or pushing it away from populated areas. Some of the methods proposed were aimed at reducing the solar radiation below the level required for photochemical reactions. Most schemes required a great amount of energy—many times more than was available then or now.

Punching a hole in the inversion

According to the article in *Science,* one suggestion was to "punch a hole" in the inversion layer. This could be done by warming the air below the inversion to a temperature higher than that at the inversion top. To do that it would take millions of tons of oil burned at 100% efficiency to produce the necessary amount of heat. However, the process of heating itself would create motion of the air which would bring new air and ozone in over the basin.

As heating that massive an amount of air was not feasible, it was then suggested that the heating be confined to a relatively small area. The hole it would "punch" in the inversion, hopefully, would enable all the pollutants in the basin to pass upward. However, whether one attempted to get the air to pass through a small hole in the inversion or to eliminate the inversion, the amount of energy required would be enormous. Therefore, it was not a practical solution.

Erecting 2000-foot high stacks

Another idea was to build a stack, or even several stacks, up to the top of the inversion. All the air which one would want to have penetrate the inversion would still have to be heated to the temperature of the inversion top, a practical impossibility. Again, ozone would be brought down from above to compensate for the upward motion of the air through the stacks.

Burning all rubbish in the county in one incinerator

By burning all the combustible rubbish in Los Angeles County in a single large incinerator, according to one suggestion, adequate heat could be engendered to cause the pollutants from it to penetrate the inversion. It would also carry with it a large amount of entrained air containing other pollutants. Unfortunately, the total amount of refuse collected in one day would be enough to only eliminate the inversion over an area of a few square miles, too small an area to make much of a difference. Again, as more ozone from aloft might be brought down, it would cancel out any gain.

To make such a plan somewhat workable, huge stacks would have to be erected up to heights of 2000 feet or more which would pose an aviation threat. Another suggestion was to have the stacks take the form of pipes going up the slope of a mountain. Thus, even more power would be required.

It's important to note that if one were to eliminate the inversion by heating the air over the basin, temperatures in the area would soar to 100 degrees or more during only average inversion conditions, creating a climate similar to desert communities during summer months.

Blowing smog away by installing large fans

Another proposal was to move the polluted air out of the basin. The amount of air that would need to be moved would weigh many millions of tons, far more than the weight of all the steel produced in the United States in a year (during the 1950s).

Moving large masses of air is difficult because it's not only hard to get hold of it but you would have to displace an equal volume of air in the place you moved it to. Fans or other propelling apparatus would have to be installed. The energy required to keep the air moving over a flat area the size of Los Angeles would be more than the amount of electricity produced by 12 Hoover Dams. There would be the problem of overcoming surface friction, as well. More than 4000 engines of 5000-horsepower would be required to maintain a 9-mile-per-hour wind over a relatively small area. It would take at least 1000 tons of fuel oil per hour to keep the fans operating. Another drawback to the fan scheme is the fact that a movement of air would increase the sea air and its natural pollutants.

Building tunnels through the mountains

Another unworkable plan was to build a tunnel through the mountains. However, if the tunnel were 100 feet in diameter and the air moved through it at 100 miles per hour, the amount passing through would amount to only a small percentage of air that would have to be moved. Much more would have to be moved to make a difference. Even 50 such tunnels wouldn't produce a noticeable effect on man-made smog and it might make the natural smog worse.

Installing fans to blow air vertically

Several proposals were made for using ground-based fans to blow the smoggy air upward. There were also a few plans to blow fresh air downward from above the inversion base using hovering helicopters. By blowing the warmer air down from the inversion layer to mix with the air near the ground, the mixture would be warmer, and might thus rise. All of the schemes ignored the fact that most ozone comes from aloft and thus would be increased. Considering what is known about natural smog, these plans were not feasible.

Natural sources of energy

Various proposals to use solar energy have been made. Suggestions include:

- Painting the roofs of buildings black and white in a checker-board pattern in alternate city blocks to promote convection.
- Paving large areas of the basin with black asphalt to eliminate the inversion.
- Introducing black carbon dust into the air over the basin at low levels so that the air in contact with it would be heated and rise

like balloons through the inversion.

- Using mirrors or lenses to concentrate the sunshine.

The above proposals are unworkable because they would bring more ozone to the ground.

Using solar energy

The use of solar energy to create greater horizontal movement of the air was another suggestion. Proponents of this plan believed that if the breeze from off-shore started earlier or was stronger, the smog would not reach such high concentrations or it would be moved out quickly.

One way to accomplish this would be to clear the brush from the larger canyons that extend into the mountains and then pave them. It would decrease the frictional resistance to flow and increase temperatures. The winds would thus start sooner and the air would move faster.

This was one suggestion that had very few supporters. The esthetic and ecological effects of stripping away large amounts of vegetation and paving wilderness areas would undoubtedly have created an outcry from many different sectors of the population.

Fog nozzles in the mountains

In a proposal to the Air Pollution Foundation, it was suggested that a string of fog nozzles be installed along the mountains at the 2000-foot contour to saturate the air. That would create a downward flow of air at night which would increase the land breeze and diminish the sea breeze during the day. It would, however, require using from 6,000 to 11,500 acre feet of water per day which, in the 1950s, was 5 to 10 times the amount of water consumed in Los Angeles each day.

As this wasn't possible in an area of water scarcity, the use of sea water was proposed. There were several concerns about this plan. One was the cost of pumping huge quantities of water from the sea to 2000 foot elevations. Another was the problem of designing fog nozzles that would not become corroded and clogged by the sea water. Still another problem was the effect of introducing such a large amount of salt from the water into the air. Not only would it reduce visibility but the corrosive action of the air-borne salt would be greatly damaging to vegetation, equipment, automobiles and other painted surfaces.

Among the other suggestions were to use water sprays to wash and cool the smoggy air but again this would require too much water. There was also a plan to reduce the sunlight by using smoke generators along the western coast of the basin to produce small droplets of oil which would scatter and reduce the intensity of the sunlight. However, this might leave a blanket of oil.

These proposals are not workable because they require enormous amounts of energy to implement, and most are based on the wrong theories. There was however, one fairly good idea which came out of that era in the 1950s.

In a small article which appeared in *Science Digest* in December, 1956, it was suggested that belts of trees be planted around industrial areas as one method of helping combat the air pollution. This proposal was made by a panel of experts at the American Chemical Society's 130th national meeting. The article pointed out that tree belts as much as 30 miles long and 3 miles wide had been planted around manufacturing districts in England to cut down on the pollution from factory smoke and toxic industrial emissions. According to the article, the trees "change prevailing wind patterns and create up-drafts around manufacturing districts to carry pollution by-products away from nearby areas."

Trees can take man-made contaminants and natural

ozone out of the air. In Los Angeles, however, we can't change the prevailing wind pattern. The plan is also impractical because of the space that would be required for the trees. There simply isn't enough land available around the factories in many areas to accommodate the numbers of trees that would be necessary.

What can be done?

What can we do about the air we breathe? Whenever possible, we can move away from an area that has an air pollution problem. We can stay indoors and not engage in exercise on smoggy days.

We can quit smoking and not allow anyone near us to smoke. Smoking is probably the most destructive thing humans can do in terms of health costs. It harms not only the smokers, but those around them. The major problem in overcoming this form of air pollution is that people become addicted to tobacco whereas no one gets addicted to smog.

Parents need to be aware of the hazards of secondhand smoke on their children's health and insist that no one smokes around them. About 6,200 children die every year as a result of their parents' smoking. It is linked to sudden infant death syndrome, lung infections and burns among other things, according to a study of the University of Wisconsin Medical School that was reported in the *Archives of Pediatrics and Adolescent Medicine.* The study also estimates that 5.4 million other children are made ill by parental smoking.

At this time there isn't a great deal that can be done about natural air pollution. However, any service which would alert citizens of an impending increase in ozone levels is beneficial. Such an alert would include natural ozone, nitrogen dioxide and particulates. To protect ourselves we might have to go into a building that has filters which remove many of these pollutants.

In the meantime, it would be wise to examine the local contaminants going into the air. The human-caused type of smog attacks can be stopped by taking several actions.

Industrial plants could install filters or other devices to reduce contaminating discharges. Plants that emit any poisonous fumes could be required to relocate to unpopulated areas. They should never be situated in valleys or areas where immobile air allows high concentrations of potentially toxic emissions to become trapped and held close to the ground by severe temperature inversions. If industries that pollute are located near a populated area, there should be a warning system that alerts the community when there's a weather pattern that may create a smog attack. And the industries should shut down during the period that they might cause air pollution.

We have no power over air pollution that is created by such natural occurrences as volcanic eruptions, earthquakes or meteors striking the earth. We can't do anything about the sulfur that emanates from land bogs and oceans or the nitrogen dioxide and ozone caused by lightning.

There is much we ***can*** do about the toxic substances that mankind is putting into the atmosphere, however, and there can be no doubt that we must maintain constant vigilance if we are to preserve the quality of the very air that supports life on earth.

Glossary

Advection
Horizontal motion of air

Aerosol
A suspension of solid or liquid particles in a gas, as smoke or fog

Ambient
Completely surrounding; encompassing

Atmosphere
The gaseous envelope surrounding the earth; the air

Basin
A hollow or depression in the earth's surface

Catalytic converter
An automotive antipollution device that changes carbon monoxide to carbon dioxide

Convection
Vertical motion of air

Eddy
A current at variance with the main current of air

Effluent
Flowing out or forth; waste that is discharged into the air or into a body of water

Emission
Something emitted; a discharge

High pressure zone
Large area where air pressure is high and air is sinking

Inversion
A reversal of the normal temperature condition; temperature rising with altitude rather than falling

Low-pressure system
Having or resulting from a low atmospheric pressure

Meteorology
The science dealing with the atmosphere, weather and climate

Micro-weather front
Boundary between different types of air formed due to local conditions and not associated with a storm

Offshore
Off or away from the shore; moving or tending away from the shore toward or into a body of water

Onshore
Onto or in the direction of the shore from a body of water

Particulate
Minute portions or particles suspended in the atmosphere, esp. pollutants

PPM
Parts per million in the air or in anything

Precursor emissions
Discharges that precede air pollution conditions

Stratosphere
The region of the upper atmosphere extending upward from the tropopause to about 30 miles above the earth, characterized by little vertical change in temperature

Subside
To sink to a low or lower level

Tropopause
The boundary, or transitional layer, between the troposphere and the stratosphere

Troposphere
The lowest layer of the atmosphere, 6 miles high in some areas and as much as 12 miles high in others, within which there is a steady drop in temperature with increasing altitude and within which nearly all clouds and weather conditions occur

Turbulence
Irregular motion of air or any substance

Updraft
The movement upward of air or other gas

Weather sink
An area of marked and persistent sinking of air from aloft

Wind direction
The direction from which the air is moving

Wind velocity
Movement of air measured in miles or knots per hour

Bibliography

Anfossi D., S. Sandroni & S. Viarengo. "Tropospheric Ozone in the Nineteenth Century: The Moncaliere Series." *Journal of Geophysical Research,* 96, 17349-17352 (1991)

American Automobile Association. "Clearing the Air: A Report on Emission Trends in Selected U.S. Cities." September 1994

Antó, Josep M. and Jordi Sunyer. Institut Municipal d'Investigaceó Mèdica, (IMAS), Universitat Autónoma de Barcelona, Barcelona, Spain. "Nitrogen Dioxide and Allergic Asthma: Starting to Clarify an Obscure Association." *Lancet,* 402-403, Vol. 345, February 18, 1995

Automobile Club of Southern California. "A Report on the Automobile and Clean Air." May 1994

Bartel, A. W. and J. W. Temple. "Ozone in Los Angeles and Surrounding Areas." *Industrial and Engineering Chemistry.* Vol. 44, No. 4, 857-861 (1952)

Blewett, Stephen E. "The Blewett Report: SMOG, What is It? Why is It? A Scientific Analysis of One of California's Most Serious Problems." *Fortnight,* Vol. 17, No. 11. December 1, 1954

Blewett, Stephen E. and Lloyd S. Davis. "Toxic Hazes Composed of Ozone and Oxides of Nitrogen of Natural Origin and Their Relation to the World Wide Air Pollution Problem." Ozone Research Group Inc. (1993)

Bojkov, R. D. "Surface Ozone During the Second Half of the Nineteenth Century." *Journal of Climatology and Applied Meteorology,* 25, 345-352 (1986)

Brauer, Michael and Susan M. Kennedy. Occupational Hygiene Program and Department of Medicine, University of British Columbia, Vancouver, BC, Canada. "Gas Stoves and Respiratory Health." *Lancet,* 412, Vol. 347. February 17, 1996

California Air Resources Board. "Stratospheric Ozone Reaches Ground in the San Francisco Bay Area Basin." *California Air Quality Data* Vol. 5, No. 1, p.7. Jan-Mar, 1973.

Chameides, W. L. et. al. "Net Ozone Photochemical Production Over the Eastern and Central North Pacific as Inferred from GTE/CITE 1 Observation During Fall 1983." *Journal of Geophysical Research* 92, 2131-2152 (1987)

Chapman, S. "A Theory of Upper-atmospheric Ozone." *Royal Meteorological Society,* 3, 125-193 (1930)

Craig, Richard A. *The Observations and Photochemistry of Atmospheric Ozone and Their Meteorological Significance.* Meteorological Monographs, Vol. I, No. 2. Boston, Massachusetts: American Meteorological Society, pp. 34-35; 1950

Davis, Lloyd S. "The Urban Ozone Ambiguity. Is Los Angeles Type 'Smog' of Natural Origin?" Ozone Research Group Inc.; July 1996

Devalia, J.L., Rusznak, M.J. Herdman, C.J. Trigg, H. Tarraf, R.J. Davies. "Effect of Nitrogen Dioxide and Sulphur Dioxide on Airway Response of Mild Asthmatic Patients to Allergen Inhalation." *Lancet,* 1668-71, Vol. 344. December 17, 1994

Dobson, G.M.B. *Exploring the Atmosphere.* Oxford University Press, 1968

Ehmert, Alfred. "Gleichzeitige Messungen des Ozongehaltes Bodennaher Luft an mehrenen Stationen mit einem einfachen absoluten Verfahren" *Journal of Atmospheric and Terrestrial Physics.* London 2(3) 1952, pp. 189-195; *Abstracts and Bibliography,* American Metro Society 3.10-70

Fabian, P. and P.G. Pruchniewicz. "Meriodional Distribution of Ozone in the Troposphere and its Seasonal Variations" *Journal of Geophysical Research,* 82: 5897-5906 (1977)

Federal Reserve Bank of Chicago. *Cost Effective Control of Urban Smog.* Papers presented at a conference sponsored by Workshop on Market-based Approaches to Environmental Policy, Federal Reserve Bank of Chicago and Chicago Council on Foreign Relations. November 1993

Franzblau, Edward and Carl J. Popp. "Nitrogen Oxides Produced from Lightning" *Journal of Geophysical Research,* 94; 11089-11104. August 20, 1989

Geological Survey. "Geology and Origin of the Chilean Nitrate Deposits." Geological Survey Professional Paper 1188. USGPO, Washington: 1981

Haagen-Smit, A.J. "The Air Pollution Problem in Los Angeles" *Engineering & Science,* 14, 7-13 (1950)

King, Clarence. *Mountaineering in Sierra Nevada.* New York: Charles Scribner's Sons, 1915

Krupnick, Alan J. and Paul R. Portney. "Cleaning Up Smog: Costs vs. Benefits" *Consumers Research,* 23-27; August 1991

Lents, James M. and William J. Kelly. "Clearing the Air in Los Angeles" *Scientific American,* 32-39; October 1993

Linvill, D.F., W.J. Hooker and A.B. Olson. "Ozone in Michigan Environment 1876-1880" *Monthly Weather Review,* 108; 1883-1891. (1980)

McKernan, Thomas. "The Facts are Clear" *Avenues,* Auto Club of So. Calif., September/October 1994

National Research Council. *Rethinking the Ozone Problem in Urban and Regional Air Pollution.* National Academy Press, Washington DC: 1991

Sax, N.I. *Handbook of Dangerous Materials.* New York: Reinhold Publishing Corp., 1951

Stewart, F.S. *Result of Quantitative Analysis for Oxides of Nitrogen (Calculated to NO_2) in Brown Clouds and Hazes.* Los Angeles: F.S. Stewart Associates, 1954

Tunnicliffe, W.S., P.S. Burge, J.G. Ayres. "Effect of Domestic Concentrations of Nitrogen Dioxide on Airway Responses to Inhaled Allergen in Asthmatic Patients. *Lancet,* 1733-36, Vol. 344, December

ORDER FORM

What's in the Air

Natural and Man-made Air Pollution

ISBN 0-9640565-2-6 ★ $11.95

✉ To order by mail, send check or money order (no cash or CODs)
payable to: Seaview Publishing
P.O. Box 2625
Ventura, CA 93001-2625

Please send me ____ copy(s) of ***What's in the Air*** @ $11.95 each

I am enclosing $________

Plus $3.50 per book for postage & handling $________

California residents add 7.5% sales tax $________

Total amount enclosed $________

Name ______________________________

Address ______________________________

City ______________ State _____ Zip __________

Telephone: (____) ______________

Please check one or more of the following.
This book(s) is being purchased for:

☐ Bookstore ☐ Public library ☐ College/University library

☐ School: ☐ Grades 1-6 ☐ Grades 7-12 ☐ Community College

☐ Individual; age: ________

☐ Other ______________________________

Prices and availability subject to change without notice.
Valid in U.S. only.